# STAR TRAILS
## Navajo

A Different Way to Look at the Night Sky

Other books by the author:
UWHARRIE LAKES REGION TRAIL GUIDE


Front cover:
Photo by Dan Heller, Death Valley; http://www.danheller.com/


Back Cover:
Photo by Don Childrey, Mt. Rogers NRA, VA
Illustration by Nickola Dudley

ISBN 1441415203
EAN-13 9781441415202

# STAR TRAILS
## Navajo

A Different Way to Look at the Night Sky

By Don Childrey

Illustrated by Nickola Dudley

After stuffing ourselves with another well-deserved gourmet trail meal, Mr. Bill and I liked to kick back and relax with our friends on some exposed mountain ridgetop. When we were lucky, as we often were, the setting sun would paint the sky for us in grand, glowing strokes of orange and blue and purple, before it slipped beyond the horizon and left us to enjoy the night. As the fading glow disappeared in the west, we would lie back on the ground and watch the stars begin to pop out of the velvet sky above. One of our favorite things to do at times like this was to count the stars that moved among the others. We knew they were really satellites, or the space station, or even the space shuttle. But to us, they were simply another beautiful part of the amazing wilderness experiences we enjoyed so much.

On a good night, we would see 20 or more of these walking stars. Mr. Bill was great at spotting them. Many times he and I would outlast everyone else and stay out until the sky was ablaze with twinkling lights. On September 21, 2002, Mr. Bill left us, perhaps to begin his own journey on the Pathway of Departed Spirits. He took his last terrestrial steps in the company of friends, while hiking on the Appalachian Trail after a day of trail maintenance, something he dearly loved. We spread his ashes on the mountain breezes, which no doubt carried him back to places he cherished and on to adventures we can't even imagine. Whenever I find myself lying back, watching the walking stars of early evening, I can't help but think that one of those lights is probably that big-hearted sparkle in Mr. Bill's eye, still shining brightly, as he explores the star trails himself.

Dedicated to Bill Medlin
1948 - 2002

# Star Trails - Navajo

# Acknowledgements

This book would not have been possible without the support and help
of several friends and family: Sandy, my wife, who patiently supported me
and let me work on this book over the last seven years; Nickola Dudley,
who created the illustrations for the Navajo story and the cover; Michele
Jozwiak, who provided expert advice on the cover design and typography
matters;

I'd also like to thank the many authors whose whose works provided
the material necessary for me to put this guide together.

# Introduction

A Fox scrambled up the rock outcropping with the endless energy of a budding teenager and lay back in the shallow depression on top. He tried to conceal himself as much as possible. Behind him, somewhere in the darkness that lay over the bald gap between Pine Mountain and Wilburn Ridge, were the Hounds. They would be searching for the Foxes by now.

Earlier that day, the whole Boy Scout troop had trudged its way to the top of the mountain after spending the night at Old Orchard Shelter. Their route followed the rocky stretch of Appalachian Trail that snaked up the wooded north slope of Pine Mountain. The moss-covered logs and ferns under the open evergreen canopy were reminiscent of school book drawings of pre-historic dinosaur habitats. The sight of a lumbering triceratops off in the trees wouldn't have surprised the boys at all. The lush, moist environment just seemed right for something unusual.

When these bearers of over-stuffed packs and sagging sleeping bags broke out of the woods near the top of the mountain, the scenery changed dramatically. Before them lay a broad rounded ridge top. Instead of mossy trees, knee-high yellowed grasses dominated this landscape. Scattered outcroppings of grayish-purple rock thrust upwards into a wind that never seemed to stop. Shoulder high thickets of rhododendron and mountain laurel were fighting their way up from the more sheltered sides of the mountain, with limited success.

A well-worn footpath wound along the ridge, passing in and out of the laurel thickets. A mile's hike along the ridge brought the boys to a bald gap. They set up camp under some stunted spruce trees at the base of a small knob. Despite the complaints of fatigue that had been heard on the hike up, and having to fix dinner and clean up, the boys were ready to romp once the sun set and darkness slid over the Virginia highlands.

Having been a Hound himself many times, the Fox lay still on the rock, confident that the erratic beam of the Hounds' flashlights would only search around the base of the rock he was on. Even if they did accidentally stray a beam of light upwards, the slight depression he was in should help

hide him from view. The younger, inexperienced Hounds would be night-blinded by their own lights anyway, seeing only what was  in the small tunnel of their flashlight beams. He knew from experience that to see in the dark, you had to be in the dark. The night world is a soup of grays and blacks, but after your eyes adjust, it is possible to move around without the telltale splash of light that could give away a Fox's path to his hiding place.

Once on the rock, his strategy was simply to wait. Above him the sky was filled with stars. Many more than he normally saw back home. And the sky here seemed bigger too. Being on top of the mountain, near the highest point in the state, he could see most of the distant horizon, something that was nearly impossible to do in the rolling Piedmont hills where he lived.

He wondered how many stars were out there. There seemed to be thousands. He also wondered which ones belonged together in the different constellations. Although he scanned as much of the sky as he could see from his hiding place, he wasn't able to recognize any of the patterns, not that he knew many to start with. High overhead were four bright stars that formed a nearly perfect square. That was probably part of one of the constellations. Not far from them, a little further to the west, were several stars that seemed to form a large cross in the sky. A little to the north of the square were five bright stars that looked like a squashed letter M. Lower down in the eastern sky was a tight cluster of stars that seemed to have a haze around them. Near the horizon he thought he saw an airplane's lights. It flashed red and green and white. Watching it for a while, he noticed that the flashing was not regular like a plane's lights. It varied irregularly in color and brightness, but it never moved, like a plane would have.

Nothing he saw really reminded him of a dipper, one of the few things he thought he might recognize if he saw it. There wasn't even an unusually bright star in the north, where he thought the north star should be. Maybe he'd try to find some kind of guide to the constellations and bring it up here next time. That would make this jumble of stars more interesting, especially if he could read the stories that went along with the constellations.

A shiver ran up his spine. The cold rock beneath him was sucking the heat from his back and a cool wind was whisking away the rest. He turned around to gaze over the gap, looking for the bouncing lights of the Hounds. There were several lights bunched together just beyond the glow of the campfire and one or two moving slowly near the edge of the laurel thickets, but none nearby. Someone started yelling "Time's up!"

The Fox climbed part of the way down the rock and then jumped the rest of the way. He quietly circled around to the far side of the campfire, wanting to appear out of the darkness from a different direction. It wouldn't be wise to give away such a good hiding place!

Growing up as a Boy Scout in central North Carolina, I was blessed with the opportunity to spend many nights out under the stars while camping and backpacking. These trips gave me a chance to look up at truly dark night skies. In the city where I lived, light pollution from streetlights and houses and parking lots blotted out all but the brightest stars. But in the backcountry, where the only other glow came from fireflies and flashlights, the skies were filled with sparkling points of light. Out there, I couldn't help but wonder about the countless stars I saw twinkling above me.

As time passed and I read more about the constellations, I could sometimes recall the names of a few of them whenever I had a chance to look up at the night sky. Every now and then I was able to pick out one of the better-known groupings of stars. Eventually I learned what the Big Dipper looked like, and that the three bright stars in a row were Orion's belt. Sometimes I could pick out the Little Dipper.

As I grew older and spent more nights in the backcountry, my curiosity about the night sky continued. Sitting out under the stars, my friends and I would point out the constellations we knew and wonder aloud about the stories behind the figures in the sky.

Usually that's as far as it went. But eventually I found enough motivation to search for even more information on the stars and the stories. I finally bought a field guide to the stars, I think it was a Peterson's Field Guide. It helped me learn more about what I was looking at and how to locate and identify other constellations. Several months later a friend gave me another book about stars.

In reading through these books, I picked up bits and pieces about each constellation's namesake, most of which came from ancient Greek mythology. The Greek myths seemed to offer a little adventure, some entertainment, to add to the simple awe of stargazing. But those field guides focused more on the science of the stars. They only provided sketchy details about the legends. That simply wasn't enough to satisfy my curiosity.

I began researching the Greek legends behind the constellations. At the time, the Internet was just gaining popularity and I eagerly used it as a research tool. Bits and pieces of a few interesting stories turned up, but I had a hard time finding a coherent story line. I was mostly coming up with brief, unrelated stories. I couldn't find a complete set of stories that included more than a few constellations. To complicate matters, a lot of the literature began with phrases like "In one story…" or "One version of the myth says…" Apparently there were numerous versions of the myths floating around out there. I began to wonder if it was even possible to find one story that linked a majority of the constellations together.

As my search continued, I ran across a web page that mentioned sky tales from Native American cultures. They were on a site created by Paula Giese, in South Dakota. Shortly after finding her web page, I ran across a message saying she had just passed away. I had just missed a chance to speak with her in person, but her web pages had already spoken to me.

Prior to seeing Ms. Giese's website, it never even occurred to me that different cultures might have their own sets of constellations and star lore. After reading a few of the Native American star stories, I was captivated. To me they were far more interesting and relevant than ancient Greek legends.

Further research revealed a familiar problem: information about Native American star lore was scarce. Instead of being discouraged, I took on the challenge. Since these stories were so little known and hard to find, I decided to write a book about Native American star lore. I wanted this book to serve as a practical guide to the night sky, one featuring constellations and stories from Native American cultures instead of Greek mythology.

As I continued my search for information about Native American star lore, I found some interesting books describing the cosmologies of different cultures. These books explored each culture's ideas about the universe and how these ideas were expressed in their mythology and star lore. Edwin C. Krupp presented an good overview of tales about the sky in his book, Beyond the Blue Horizon. He suggested the main function of sky tales was to reveal how the cosmos works and what each person's place in the universe might be.

When you look at the night sky, I mean really stop and take the time to look at it carefully, you can begin to sense just how vast and majestic it really is. If you happen to be stargazing by yourself, it's easy to feel small and insignificant, alone out there in the dark. Especially when you try to comprehend the endless spread of stars and empty space before you. The immense scale and mystery of the night sky can be overwhelming.

Telling stories about celestial objects and events gives us a way to link ourselves to the workings of the universe. Nearly every culture has sky tales in one form or another. Some cultures have incorporated these tales into their religions. Others have made them part of popular entertainment, as with the contemporary Star Trek tales. Listening to these stories, we strengthen our realization that we are part of this vast universe in motion. Recognizing this connection gives us a more secure sense of who we are and how we fit into the world around us.

Early Native Americans paid more attention to the celestial patterns and cycles happening above them than most of us do now. For them, observing the sky was more than just a way to satisfy their curiosity. Krupp noted that sky tales enabled them to comprehend their environment and interact with it in practical ways. Careful observation of the patterns and rhythms in the sky helped them understand the patterns and rhythms found elsewhere in their world.

Many early Native Americans depended heavily on celestial guides for the proper timing of food production and gathering. Recurring celestial events, like the rising of a certain group of stars, were used by the Navajo to mark the proper time for planting corn and by the Tlingit to predict the arrival of migrating salmon.

The stars' repeating pattern of motion allowed the watchers to accurately mark the passage of time during the night and the passage of seasons during the year. With its apparent lack of motion, the North Star, now known as Polaris, was used and a landmark to help travelers orient themselves correctly.

Paying attention to heavenly cycles allowed these people to work towards achieving a balance with the environment in which they lived, an environment which could provide the elements necessary for their survival. I see the Navajo principle of reciprocity in such a relationship: if the people gave proper care and respect to the world around them, they could expect that world to take care of them by providing for their needs. Crops planted at the right time would be more likely to produce good harvests.

Sustenance and ritual were woven together tightly in the fabric of many Native Americans' daily lives. Certain rituals were often called for in response to heavenly events.

But while all members of a tribe were involved with the rituals to some extent, much of the knowledge of the heavens was usually kept by a select few like the priests of the Skidi Pawnee or the Navajo chanters and stargazers. The average member of the tribe only knew a small part of this collection of celestial information.

Preserving the knowledge gained from hundreds of years of observing the sky without a written language was not easy. Keepers of celestial knowledge managed this feat by incorporating some of the information into the sky tales of their culture. These legends and myths were passed on orally, as stories and ritual songs and recitations. Long winter nights were the traditional storytelling times, although some stories were told year round.

In addition to the preservation of information, storytelling also served as an effective way to teach proper cultural values to members of the tribe. Repeated tellings of the stories helped maintain awareness of these values. Tying the stories to ever-present celestial objects enabled the heavens to provide constant reminders of these values. Thus the heavens were preserved in the stories and the stories were preserved in the heavens – another example of the Navajo principle of reciprocity.

In her book Earth is My Mother, Sky is My Father, Trudy Griffin-Pierce said, "The power and strength of myths lies in their ability to reach through time with themes that are both universal and eternal." These myths were passed on from generation to generation, keeping the knowledge and cultural values of the tribe alive.

There were some drawbacks to the oral-based system. Details in the celestial stories often varied from tribe to tribe and even from band to band within the same tribe. The oral nature of the stories also allowed them to change over time. Details were altered to be more relevant to a group's current conditions and location. Influences from contact with other tribes were often added to the stories.

When Europeans began to settle in North America, conditions changed drastically for the indigenous peoples. The pace of change was fast. Even the extraordinary adaptability of Native Americans could not keep up. Many tribes were relocated away from lands that were sacred to them. Forcible efforts were made to assimilate them into Euroamerican culture. Many of the stories and legends that had preserved their accumulated cultural and celestial knowledge were forgotten in the turmoil.

By the time literate people began to recognize the value of the Native American stories and make efforts to record them, a significant amount had already been lost. In the case of the Skidi Pawnee star lore, some of the primary informants for the first written records had not performed the sacred rituals in over 30 years. It is impossible to say whether any story we have now is recorded exactly the same as it was told in the past. When you compare several historical documents referring to the same story, contradictory details are common.

Differences in details existed in these stories before Europeans arrived in North America. Cultural assimilation nearly wiped out the oral traditions that had preserved these stories for generations. Perhaps the true value of these stories was not found within their details. Instead, their importance came from the way these stories were woven into the context of the peoples' daily lives and in turn helped defined their place in the fabric of the universe. The stories enabled the people who heard them to better appreciate and coexist with the world around them.

After months of reading and research, I decided to focus this book on the star lore of the Navajo people. This tribe had a fairly complex and detailed cosmological system. But most importantly, a relatively ample amount of detailed information about this tribe's star lore, thanks to the efforts of many people, like Berard Haile and Trudy Griffin-Pierce.

In order to create a more fluid presentation for this book, I have taken a few liberties in retelling the stories I learned. Perhaps this is my way of adapting them to fit the circumstances of my time.

I found several slightly different versions of the Navajo star creation story. They differ based upon which ceremonial tradition they come from. I chose to present a creation story from just one, the Nightway tradition.

I do not intend for this work to serve as a documentary of Navajo star lore, although I have tried to represent information accurately where possible. Instead, this work should be viewed as an interesting and perhaps enlightening introduction to Native American star lore. If these stories spark your curiosity, I urge you to follow up on your interest and learn more about Native American star lore and culture.

Why the title Star Trails? Time lapse pictures of the night sky reveal the circular paths that stars follow across our sky. In astronomy circles, these are known as "star trails". I thought this name would make a perfect title for my book because it makes me think of backcountry adventures and ties in with the practical knowledge that Native Americans gained from observing the travels of the stars.

My original thoughts were to write this book for people like myself, who often find themselves camping out under dark backcountry skies. As it turns out, anyone willing to venture outside at night and get away from the blinding influence of artificial light can use this book. I hope it will help you learn more about the night sky. Perhaps you will even begin to see the heavens as Native American eyes might have in the past.

# Visual Astronomy

### Observing the night sky

Early Native Americans watched the skies and learned a lot about the heavenly bodies without the aid of telescopes or binoculars. There is much to be seen and appreciated in the night sky if you take the time to look. Just looking with your naked eyes you can see about two thousand stars, five planets, and the phases of Earth's moon. You can also see unusual phenomena like meteors, comets, and eclipses. Most of what you can see without aid lies within the disk of our own galaxy. The furthest object you might be able to see is the Andromeda Galaxy, lying about 2 million light years away.

Observing the sky night after night enables you to recognize the rhythmic cycles of celestial objects. The main hurdle you have to overcome is leaving our habitual dependency on artificial light and indoor environments. We must go outside, into the dark, to look up at the night sky. If we can get past these mental obstacles, there is literally a whole universe of beautiful things to see in the night sky.

### Viewing hints

Before you go out to look at the sky, here are a few bits of advice to help make your viewing more enjoyable. First, it's cold outdoors at night. Even in summertime, temperatures drop at night and it's easy to get chilled. This is especially true if you're sitting still, as you will be when stargazing. Dress warmly or bring a blanket to wrap around you and/or the person next to you. This is extremely important if you've persuaded a spouse or friend to come along despite their lower level of enthusiasm. You'll have the fire of curiosity to help warm you. They may need the blanket. Neither of you will enjoy stargazing very much if one of you is shivering and distracted.

Looking at the stars obviously involves a lot of looking up. Our necks weren't made for looking skyward for any length of time. Stargazing can give you a serious crick in the neck if you aren't prepared. Laying back in

a reclining lounge or deck chair is a good way to reduce neck stress. Lying on a mat or blanket on the ground is another option. A pillow or headrest of some kind can be useful too. Regardless of how you do it, anything is better than simply trying to endure the pain of looking up for long periods of time. If you're not comfortable, you won't spend much time looking at the stars.

## Factors affecting viewing

When you go out to look at the sky, several factors can influence how much you will see. Weather obviously plays a significant role in what stars you can or can't see. Clouds, haze and other atmospheric moisture can reduce visibility. This is especially true near the horizon where you are looking through more atmosphere than when looking straight up. The effects of atmosphere are what create the "twinkling" of stars. This twinkling is more pronounced near the horizon. You'll often see these low stars blinking or varying in brightness and color. The lower humidity of winter air tends to clear the atmosphere and allows you to see the stars better.

One primary concern is the amount of light pollution coming from streetlights or other development around you. If you have a chance to get away from artificial light sources, do so! You'll be able to see a lot more of what's in the night sky. My favorite viewing locations are places I've found after backpacking into the mountains, miles away from any development.

The moon can also significantly affect how many stars are visible. Especially when the moon is close to full, the light reflecting from it will overpower all but the brightest stars. In the *Helpful Charts and Data* chapter, you will find a series of charts indicating phases of the moon and whether or not the moon will be above the horizon during the first few hours after sunset. You can use these yearly charts to plan ahead for the best viewing opportunities.

Another factor is your eyes' adjustment to the light. After leaving a well lit place, it may take your eyes 5 to 15 minutes to adjust to the dark. The time required will vary from person to person. One way to avoid losing your adjustment is to use a red lens on your flashlight when you need to refer to this book or other star guides. Even a piece of red cellophane over a regular flashlight will work. Red light doesn't cause your pupils to constrict as much, thus preserving your night vision.

Finding a viewing location where you can see a lot of sky is important too. You can see bits of sky from almost anywhere, but until you find a spot where you can see most of the sky, you won't be able to appreciate the vastness of it. A good hilltop or mountaintop site reveals a sky so big that you can't take it all in with one look. Everyone can't get to a mountaintop, but you can try to find an open field or high vantage point from which to observe.

### Magnitude

Another factor determining visibility is the relative brightness of each star. In 129 BC, the Greek astronomer Hipparchus devised a brightness scale where the brightest stars were called "first magnitude" and the dimmest ones visible to the naked eye were called "sixth magnitude". In 1856, an English astronomer named Norman Pogson developed a more precise scale. It was based on the mathematical concept of a first magnitude star being 100 times brighter than a sixth magnitude star. These numbers representing a star's brightness are called a star's magnitude. Magnitudes range from negative numbers to positive, with the negative numbers being the brightest objects. Naked eye observers with good viewing conditions can usually see stars down to magnitude +6.0.

### The best times to observe

The best viewing usually occurs after astronomical twilight, although you will be able to see stars between sunset and the end of twilight. Astronomical twilight normally ends between sixty and ninety minutes after sunset. In the *Helpful Charts and Data* chapter are charts graphically indicating sunset, sunrise, and astronomical twilight times for the next several years. These charts also present rise and set times for the five naked eye planets. You can use these charts to see what planets should be visible at a given time.

Going one step further by bringing along binoculars will add another level of wonder and excitement to your night sky viewing. Binoculars bring thousands of additional stars into view. If conditions are favorable

you may also be able to see several galaxies and even a few nebulae. The moons of Jupiter may also be visible.

For a rough estimate of time remaining until sunset, hold up your hand at arm's length. With your fingers held horizontally, each finger's width represents approximately 15 minutes of sun travel. For example, if there are three finger widths between the sun and the average horizon, the sun will be setting in roughly 45 minutes. Likewise, finger-width estimating can be used to guess when a star or constellation will set.

## Sky orientation

The star finding maps in this book use references to compass directions to guide you to the right part of the sky. It would be helpful for you to know which way is north at your viewing site. Once you learn to recognize Polaris, the North Star (it will always be in the north), you won't necessarily need a compass.

One common celestial term is zenith. The zenith is the point in the sky directly overhead. Since the horizon is normally considered to be flat all around, the zenith will be found 90 degrees up from the theoretical horizon. Unless you are out on a wide treeless plain or a relatively high mountaintop, part of your view of the sky near the horizon will probably be obstructed.

Meridian is another common term. The meridian is an imaginary line drawn from a point on the horizon at true north, across the sky passing through the zenith, to a point on the horizon at true south. Stars and asterisms revolving around the celestial pole reach their highest points above the horizon when they cross the meridian.

Latitude is also an important term. The spherical earth is divided up in degrees from the equator to the poles. The equator is at 0 degrees latitude. The North Pole is at 90 degrees latitude north. A handy rule of thumb to remember is that Polaris, the North Star, will always be found the same number of degrees above true horizon as the latitude of your viewing location. I have drawn the star maps according to the latitude of my home in central North Carolina.

For comparison, the chart below lists the latitude of a few relevant locations:

Navajo homeland  36° 00' N
Mt. Rogers, VA  36° 30' N
Asheville, NC  35° 30' N
Atlanta, GA  34° 00' N
Greensboro, NC  36° 00' N
Raleigh, NC  35° 40' N
Richmond, VA  37° 30' N
Athens, Greece 37° 58' N
Seattle, WA 47° 34' N

### What you will see up there

Another way to increase your enjoyment while viewing the night sky is to learn a little about what you'll be seeing before you go out. Star Trails provides you with a brief introduction to objects in the night sky. I would encourage you to read further in the books listed in the bibliography to learn even more about astronomy. Star Trails is not intended to be a complete source of astronomical knowledge. It is merely an introduction to a subject I find interesting. I hope it will inspire you to seek out more information on your own.

### Moon

The moon also has its cycles, accentuated by the fact that its motion around the earth is slower than the sun's daily travels. The difference in relative position between the earth, moon and sun accounts for the various phases of the moon. The new moon occurs when the moon is between earth and the sun and the sunlit side of the moon faces away from us. The full moon occurs when the moon is on the far side of the earth from the sun. At that time, we see the fully lit side of the moon. The moon changes from new to full and back to new again roughly once a month.

Some Native American groups used the moon's cyclic patterns to set calendars. Our own modern calendar is also based on lunar months, which are actually 29.5 days long. You can see the problem that arises because twelve lunar months do not fit exactly into one solar year. Some Native Americans tried to solve this problem by adding an extra month every three years or so. Our contemporary calendar adds extra days to some months and a leap day every four years to reconcile the difference.

Each month's full moon has a special name. Many of these names originated with Native American tribes. Learn more about the full moon names in the *Helpful Charts and Data* chapter.

**Stars**

Anyone who gives the night sky more than a passing glance should notice the stars also appear to move across the night sky. Looking at them night after night, you should notice they start out further along in their journey across the sky each night. The stars in the east appear a little higher above the horizon than the night before. If you were to observe which stars were just above the horizon at dark every night, it would be one year before you saw the same ones there again. This cycle of the stars corresponds to one Earth year. For someone looking to understand the rhythms of the seasons, this makes the star cycles a convenient way to mark and predict the passage of time.

Over the course of a few years or even a few generations, the stars appear to move as a whole across the sky. In reality, the individual stars are moving in many different directions. Fortunately for us, their apparent movements across the sky are so minute; you and I probably won't notice a difference in our lifetimes.

Remember the star that does not move; the North Star? Well, it is moving. For those of us in the Northern Hemisphere, the center of sky rotation, called the "north celestial pole", lies conveniently close to the star named Polaris, commonly called the North Star. For the last several thousand years, people have been able to point to this star and say that the others rotated around it. However, a slight wobble in the earth's spin, called precession, causes this point in the night sky to move in a huge circle over the course of about 26,000 years. The north celestial pole will come closest to Polaris in 2100 AD. After that it will begin to move away. In another 13,000 years, Vega will be the new "North Star." For observers in the Southern Hemisphere, there is no star close to the South Celestial Pole at this time.

For more details on star types and colors, please see the *Helpful Charts and Data* chapter.

### Groups of stars

Certain groupings of stars are easily recognized. These groups of stars are commonly referred to as constellations. But that's not what astronomers mean when they say "constellation". Officially, a constellation is defined as a section of the sky.

At the beginning of recorded history, people in the Middle East devised some of the main constellations found in use today. They formed their constellations in the likenesses of fabled creatures and mythical heroes they imagined they could see in the patterns of stars. In 150AD, a Greek astronomer named Ptolemy recorded a list of 48 constellations. Later, various other astronomers added constellations to fill in gaps and to cover the newly explored Southern Hemisphere skies. Some of these constellations were adopted while others have fallen out of use over time.

In 1930, the International Astronomical Union divided the entire sky into 88 official constellations. Most of the easily recognizable groupings of stars do fall within the same "constellation", or area of the sky. The primary purpose for dividing the sky into sections and naming constellations has usually been to make locating celestial objects and patterns easier.

Most stars visible to the naked eye are officially named by assigning each a letter of the Greek alphabet followed by the constellation name. An example would be Alpha Centauri, the brightest star in the constellation Centaurus. The brighter stars also have proper names, most of which were designated by Arab astronomers over 2,000 years ago.

Recognizable star groupings, which may include stars from multiple constellations, are called asterisms. The Big Dipper, the Summer Triangle, and the Winter Circle are examples of asterisms.

Most early Native Americans did not recognize or use constellations as we define them today. The Navajo recognized a few asterisms which held meaning within their culture. Some of their asterisms served as timing or calendrical devices. They were not concerned with sectioning off the sky or identifying every single star. They believed a certain amount of chaos was to be expected in the universe and accepted the remaining stars as just that, the result of chaos.

## Planets

There are certain bright objects in the sky that change position quite noticeably over time. These objects are actually other planets within our solar system. Most people today do not realize that what they may refer to as the Morning and Evening Stars are actually planets. Early Native Americans did not realize they were planets either, but they did recognize them as something special. The Skidi Pawnee believed them to be major "star gods" who had powers beyond the other star gods. The unusual motions of these "wandering stars" were accounted for by the Skidi in their stories of celestial events.

Nearly all stars appear as points of light, even in the largest telescopes. But the disk shape of some planets can be seen even with binoculars and smaller telescopes.

Mercury is the closest planet to the sun and is never seen very far from it. Being so close the sun's glare makes Mercury difficult to spot.

Venus is next in the planet lineup and is the brightest planet seen from Earth. It is often visible through urban light pollution and sometimes in the daytime sky, if you know where to look. Because it is closer than Earth to the sun, Venus is only seen in the evening or morning hours.

Mars is next the next furthest planet from the sun after Earth. Mars exhibits what is called retrograde motion in its path among the stars. In layman's turns this means that at times, Mars appears to stop and move backwards in relation to the star field behind it. Mars is distinguishable by its reddish color, in contrast to the whiter colors of Saturn and Jupiter.

Strong binoculars or a small telescope are required to see the rings around Saturn. They are Saturn's most notable feature.

Jupiter is special for viewers with binoculars because of its four largest moons. With clear dark skies, you can distinguish these Jovian moons. Nightly observations will reveal their varying positions as they orbit around Jupiter.

## Other objects or events

From time to time, various other objects become visible in the night sky. These include passing comets, supernovae, and meteors as well as a host of manmade satellites orbiting our planet.

The Skidi Pawnee believed meteors to be children of their supreme being, Tirawa, and revered any meteorites they found. Learn more about when to view meteor showers in the *Helpful Charts and Data* chapter.

Early Native American observers noticed other irregularly occurring events as well. The most notable of these were eclipses. These unusual events were interpreted in various ways. Many Native Americans saw eclipses as signs of an imbalance in the world or of impending misfortune.

# Navajo Story

White sand fell to the floor in a slow, deliberate stream. A weathered thumb held lightly against a bent forefinger let the sand pass from his hand in just the right amount. Watching the old Navajo man was mesmerizing. He was forming a nearly perfect line across the floor. The white sand contrasted sharply with the underlying bed of tan colored sand that had been spread over the dirt floor of the hooghan earlier that afternoon.

The old Navajo's assistants pulled a string along another segment of the outline they were creating. After they stretched the string taut, the old man gently plucked the middle, snapping a faint depression in the sand. He followed the depression with more white sand, continuing the outline of the sand painting figure.

White is one of the four primary colors in the traditional Navajo universe. The pure white sand outlining this figure signified the white light of approaching dawn.

Exact placements and symmetry within the sand painting were vital to its effectiveness. In fact, even the hooghan they were in had been built with the door facing east. The stacked and notched logs forming the walls of this circular structure held up the low earth-covered roof. A small hole in the center of the roof allowed smoke to escape from the cooking fire that would normally have been found in the center of the floor. Like many others over the centuries, this one room hooghan usually served as a Navajo family's home; kitchen, dining room, and bedrooms all in one. Now it was being transformed into a blessed place for a healing ceremonial.

Tonight, the fire pit had been filled in with dirt and the sand painting was spreading across the floor in its place. Tradition and ritual required a hooghan's single door to face east, the direction from which all good things come. The heads of the two figures in the sand painting were also oriented towards the east. Each line and angle they drew was measured to make sure that lengths and proportions were correct.

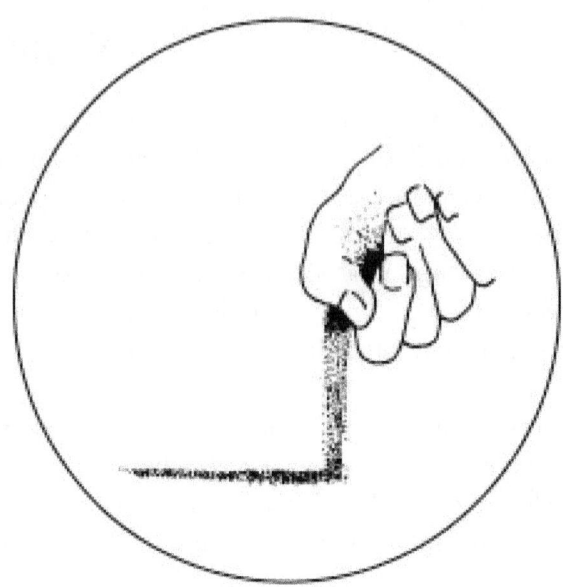

Other assistants were already at work filling in an identically shaped figure drawn next to the black one. The assistants filled this figure with bluish-gray sand made from ground turquoise. Blue represented the brilliant blue of the midday sky and was the second primary color. On this figure's belly would be drawn the four sacred plants: corn, beans, squash, and tobacco. The roots of each plant rose from a lake representing the Place of Emergence. Navajo origin legends describe how the Dine, or Navajo people, emerged into this world at the Place of Emergence.

The blue bodied figure had a yellow outline. Yellow was the third primary color of the traditional Navajo universe and represented the golden glow of twilight. When completed, the blue-bodied figure would be called Mother Earth.

Ground charcoal and white quartz were stirred together to form a jet black sand. Assistants took the black sand and began carefully filling in the body of the figure outlined in white. Black was the fourth primary color. It represented the darkness of night.

The left arm of the black-bodied figure had been drawn so it crossed over the right arm of the Mother Earth figure. The same had been done with the legs.

When the white outline had been completely filled in with black sand, the old Navajo began drawing a thicker line with white sand in a zig-zag pattern across the arms and chest of the black-bodied figure. A second zig-zag line was drawn crossing the first, forming a pattern of connected diamonds. He then started placing tiny x's and dots of white sand on the figure's black belly. This was to be Father Sky.

One of the observers asked the old man to tell the story of how the stars were placed in the sky. The old man smiled and paused for moment. He then eased back from the sand painting and sat in a more comfortable position. After gathering his thoughts for a few seconds, he began the story ...

"A long time ago, before the people emerged into the Surface world, the *diyin dine'é*, or Holy People, gathered together to create things for this world. They met in the first hooghan, specially built as a place where they could carefully think and plan for the orderly creation of the world we now live in.

When they were ready, the Holy People drew two figures on the floor of the creation hooghan, using colored sand. The heads of these figures lay towards the east. Their feet were towards the west. East is where new life and blessing comes from. One figure they called Mother Earth. The other they called Father Sky.

The Holy People sang and *nílch'i*, the Holy Wind, entered Mother Earth and Father Sky and made them alive. *Nílch'i* is the breath of life that animates all living things in the universe. We are all related because we share this common element.

The Holy People then dressed Mother Earth with various plants that would grow and provide food for the people and animals. They put storms and rain on Father Sky to provide water for the people and animals.

First Man pointed out to the group that they had been working for a long time without resting. At this time there was no order to periods of work and rest. So First Man painted Sun and Moon on Father Sky. Sun would provide heat and light to the world. He then chose He-Who-Returns-

Carrying-One-Turquoise to carry Sun across the sky. Sun's travels would divide day from night. During the day, people would have light by which to work and travel. First Man then chose He-Who-Returns-Carrying-One-Kernel-Corn to bear Moon across the sky. Moon would provide a little light to the world during the night. People were to rest during this dark time.

The Holy People agreed that dividing day and night was good. But they noticed that Father Sky was still rather dark and plain during the night, even with the light provided by the moon. While they were thinking of what to do about this, *Haashch'ééshzhiní*, Black God (or Fire God) arrived.

Black God entered the hooghan and paused just inside the doorway. Attached to his ankle was a small cluster of sparkling crystals. The other Holy People knew he had power over the energy of fire and lightning and rainbows, but they had never seen anything like these crystals before.

Without saying a word, Black God walked around the sand painting on the floor in the sunwise direction, first going to the left, around the south and west sides. He stopped on the north side. Black God stamped his foot down on the floor of the hooghan. The cluster of sparkling crystals jumped up to his knee. Again he stamped his foot on the ground. The sparkling crystals jumped up to his hip. The Holy People looked at each other and nodded approval of Black God's ability to control the interesting crystals.

Still silent, he stamped his foot down hard again. This time the glittering crystals jumped from his hip up to his shoulder. Stamping his foot one final time, Black God caused the sparkling crystals to jump to his temple.

"And there they shall stay!" he said.

The other Holy People were impressed with Black God's show of power in placing the crystals on his temple. "What are those sparkling crystals?" they asked. "They are so beautiful!"

"They are called stars," Black God answered.

One of the Holy People suggested that Black God's stars would look nice on Father Sky during the night. They discussed this and quickly agreed that Black God should be responsible for making more stars to place on Father Sky to make him more beautiful.

First Woman then spoke, "Everything in this world has been created and given its own special place. There is *hózhǫ* here, a state of balance and beauty. This order and balance can only be maintained through proper relationships among all living things, between the smallest animals and plants, even people and mountains. It is the responsibility of every living spirit to honor and maintain this balance, to seek *hozho*."

"Earth Surface People will need reminders of the rules to live by, in order to maintain balance and beauty in the world. The water is always changing, so that is not a good place to put these reminders. The earth will change too, so it is not a good place either. If these reminders are written in the sky they can be seen by all and remembered forever."

First Man nodded agreement and suggested that Black God use his stars to place these reminders on Father Sky. The other Holy People approved of these suggestions.

Black God agreed to help them do this. He pulled out a pouch that he always carried with him. It was made of the spotted skin of a fawn. Black God opened his pouch and took out a single bright crystal. He reached out over Father Sky and carefully placed the crystal in the north.

"This will be North Fire," he said. "This star will be the only one that never moves. It will guide the traveler. This world will be a place of motion. Movement is the basis of life. All other stars will move through the sky around North Fire. "

One of the Holy People suggested there should be a reminder of family and kinship responsibilities. Black God reached into his pouch and pulled out seven large crystals. They did not sparkle like North Fire. But when he laid the first one down near North Fire, it began to sparkle like the crystals on his temple. As Black God placed each of the other crystals on Father Sky, they began to sparkle too. He arranged them in a pattern that represented a man lying beside a fire.

"North Fire is the igniter star for this star figure," said Black God. "It provides the light that makes the other stars sparkle. This pattern of stars will be called *Náhookqs bika'ii*," he said. "Its name means Revolving-Male. He will circle closely around North Fire and will always be visible in the night sky."

Black God reached into his pouch again and pulled out five more large crystals. As he laid each one down on Father Sky, they too began to sparkle. Black God placed them near North Fire, across from Revolving-Male. He then arranged them in a pattern that represented a woman lying beside a fire.

"These stars will be called *Náhookos ba'áadii*," said Black God. "It means Revolving-Female. She will circle around North Fire with Revolving-Male. Together they will remind Earth Surface People of the importance of family responsibilities. Their circular path around North Fire will symbolize the shape of a hooghan, which will be the home dwelling of the Navajo. North Fire will represent the fire on the hearth in the center of the hooghan, providing warmth and life to those within. Because these two are man and woman, they will also represent First Man and First Woman. Their

position on opposite sides of North Fire symbolizes the importance of balance. They are the two halves of a whole."

The Holy People decided there should be a star pattern to remind earth people of the importance of wisdom, right thinking, and the pursuit of discipline.

Black God took more crystals out of his pouch. He began laying them on Father Sky in the east. He arranged them in a pattern resembling a man searching for something in the distance with feet braced apart and his hand shielding his eyes. Like the others, these crystals did not shine at first. Then Black God pulled out a special crystal and placed it near them. At once the crystals all lit up and began to sparkle.

"These stars will be called *Hastiin Sik'ai'i*," he said. "This name means Man-With-Legs-Apart. *Hastiin Sik'ai'i* will remind people of the importance of right thinking and concentration. Right thinking will keep their minds sharp. Man-with-Legs Apart will remind people that guidance is always there if they look for it. The search for balance will help them avoid illness and misfortune."

The Holy People agreed there should also be a pattern to remind earth people of their interconnectedness to the rest of the universe. Reaching into his pouch once more, Black God pulled out another group of crystals. He began placing them on the south side of Father Sky. Black God arranged them in the figure of a large person. Again Black God took out a special crystal and placed it near the others. Immediately they began to sparkle and shine.

"These will be called *'Átsé'etsoh*," he said. "This means First-Big-One. First-Big-One will be visible in the night sky during the growing half of the year when food is plentiful. Because First-Big-One is in the shape of a human, it will remind earth people, during their time of plenty, how closely they are related to the rest of creation. First-Big-One is a living being, just as humans are. People and stars are made of the same material as all living things."

One of the Holy People said that Earth Surface People should honor the animals that gave their life force to feed humans. Black God pulled several more stars from his pouch. As he laid them on Father Sky near First Big One, he arranged them in a pattern resembling the tracks of a rabbit. When he finished, he placed an igniter star near them so that they all began to shine.

"These shall be called *Gah heet'e'ii*," he said. "This name means Rabbit-Tracks. This honors the rabbit for his contribution as an important food source for the people. Rabbit-tracks shall be the hunter's guide. When it rises in the spring just before Dawn, it will mark the time when female antelope are bearing their young. No one shall hunt them during this time so that they may raise their young and provide another generation. When Rabbit-Tracks lies on its side at twilight, the young animals will no longer be dependent on their mothers and hunting can begin again."

The Holy People also knew that plants would be important to survival of the Earth Surface People. Black God took more crystals from his pouch and placed them on Father Sky in the west. He arranged them in a pattern resembling a thin person. "These will be called *'Atsé'ets'ózí*. This name means First-Slim-One. It will be visible in the night sky during the winter half of the year, when food becomes scarce and people grow lean. When First-Slim-One is seen in the western sky at twilight, it will be a sign to the Navajo that they should begin planting their crops of corn, beans, squash, and tobacco. First-Slim-One will be the keeper of the months. He will make sure the other stars appear in the sky at the proper times of the year.

Black God created several other patterns of stars and placed them in the sky. They included Horned Rattler, Thunder, Bear, and others. Then Black God pulled out seven small and fine crystals and carefully made a copy of the pattern of sparkling crystals on his temple. Black God placed them on Father Sky and said, "These will be called *Dilyéhé*. They will mark the proper time for ceremonials."

After he placed *Dilyéhé*, Black God reached into his pouch and pulled out thousands of tiny crystals. Spreading them across Father Sky with a long sweeping motion he said, "These stars shall be called *Yikáísdáhí*. This means Which-Awaits-The-Dawn. It will always be visible in the night sky. When Dawn approaches, Which-Awaits-The-Dawn will brighten and signal the coming of a new day."

After Black God placed Which-Awaits-The-Dawn, he started to sit down and admire their work. At this point Coyote stepped among the group of Holy People. As was his habit, Coyote had been hanging around looking for ways to play a trick or cause trouble. "What are you doing?" he said, "You didn't ask for my advice!"

Black God replied, "Look and see for yourself. Look at all the beautiful patterns we have created on Father Sky. These patterns will remind Earth Surface People of the rules to live by."

Black God started once more to sit down, crossing his legs as he always did. He was about to place his fawn skin pouch under his foot to protect it when Coyote snatched it away.

"I can help too!" Coyote said with a grin on his face. Before anyone could stop him, Coyote opened the pouch and blew the rest of the crystals over Father Sky. Thousands of stars were scattered across Father Sky in a disorderly jumble.

Coyote looked into the pouch and found one more crystal. "This one will be my star," he said. Making fun of Black God, Coyote reached far to the south and placed the star in the night sky. "The Coyote Star will only be visible a few days each year."

Black God scolded Coyote for disturbing the orderly arrangement of stars on Father Sky. "Now the night sky is full of chaos. It will be harder for Earth Surface People to find the guiding patterns we have placed for them," he said.

Coyote laughed and tossed the fawn skin pouch back to Black God, "Yes, but now the skies are beautiful!"

# Navajo Star Figures

# North Fire

North Fire was one of the first stars placed in the night sky by Black God. It is not a separate star figure in itself, but is associated with Revolving Male and Revolving Female, two star figures that circle around North Fire.

North Fire represents the fireplace hearth in the center of a hooghan, a traditional Navajo dwelling. Family ties and responsibilities were of great importance to the Navajo, so the hearth fire could be said  to lie at the center of Navajo life, both literally and figuratively.

North Fire's position in the center of the revolving sky indicates the importance of home and family. Many activities of a Navajo family happen around the hearth fire. This star functions as a constant reminder of one's family duties and responsibilities.

North Fire's unmoving position also serves as a guide to people who are traveling. It is a fixed point of reference that travelers can use to stay on course. However, the Navajo considered moving about at night to be very dangerous. Night travel exposed one to many dangers,

including natural hazards as well as evil influences.

The Navajo's North Fire is the star astronomers today know as **Polaris**.

Polaris, which means "pole star," was given this name because it lies close to the current northern celestial pole. The celestial pole is the point in the sky that lies on the axis about which our planet rotates. Because our planet spins on this axis, the sky appears to rotate around this point.

Other names for this star include North Star, Pole Star, and Star of the Sea. Since the heavens appear to rotate around Polaris, it has received special recognition in nearly all star-gazing cultures, especially those who navigate on the sea or travel over vast distances.

Even though most star stories describe this star as unmoving, this isn't quite true. A slight wobble in the earth's rotation causes the celestial pole to shift in position over time, making a full circle over the course of about 26,000 years. The apparent celestial pole will pass closest to Polaris in the year 2105, so many generations to come will see this star as the center of the night sky.

███ ✦ ███

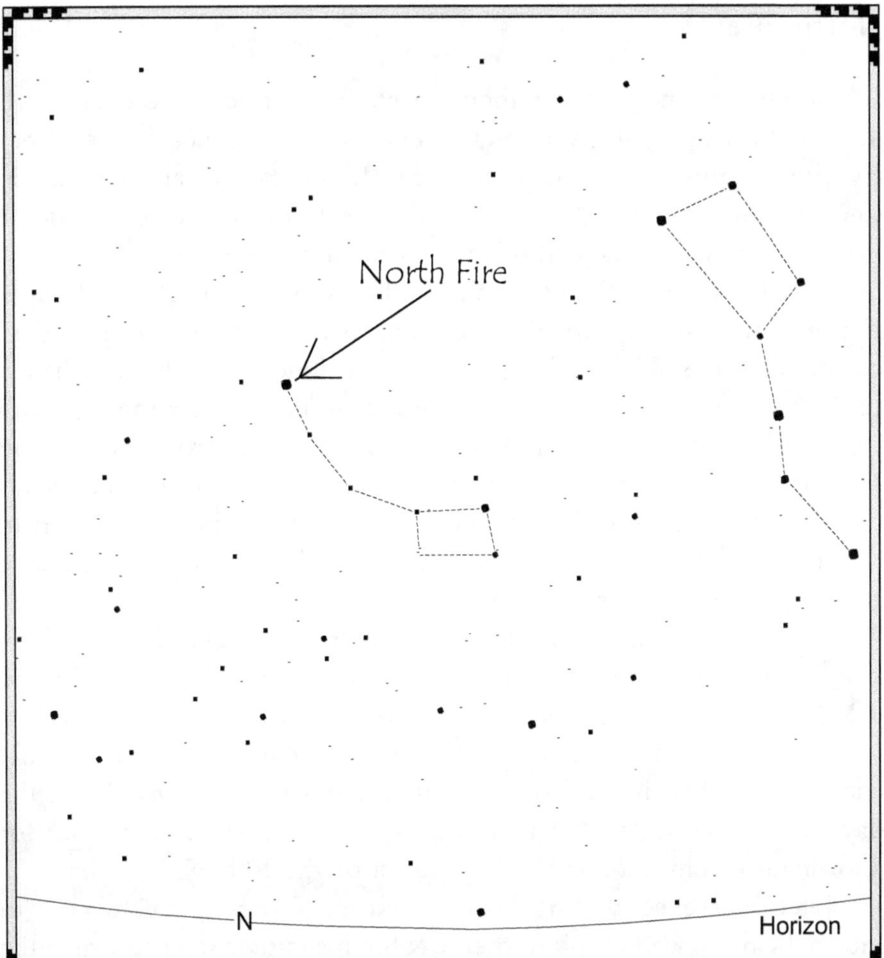

North Fire

North Fire as seen on March 21, around 7:30 PM, EST.

N

Horizon

## Finding North Fire:

Because North Fire lies close to the northern celestial pole, it doesn't seem to move at all. This star will always be found in the direction of due north.

It's height above the northern horizon will be the same as your latitude. If your latitude is 35 degrees north, like most of NC, then North Fire will be 35 degrees above the true horizon.

Remember there are 90 degrees between the true horizon and the zenith, a point directly overhead. Your apparent horizon, or what you see as the horizon, may vary if there are mountains or trees in the way.

## North Fire

**Thuban** was the pole star about 4,000 years ago. Many ancient Egyptian temples were oriented towards Thuban.

North Fire lies at the end of the upturned handle of the Little Dipper, an asterism within the constellation Ursa Minor (UMi), the Little Bear.

According to Greek mythology, the Little Bear is Arcas, son of Callisto. Jupiter seduced Callisto and she bore his son, Arcas. Jupiter's wife Juno found out and turned Callisto into a bear who ran away to hide in the forest.

After Arcas grew up, he became a hunter and encountered a bear in the forest one day. He was about to kill it, not knowing it was his mother.

Jupiter intervened, turning Arcas into a bear as well. Jupiter then grabbed them by their tails and threw them into the sky, where they could be together forever. Callisto is the Great Bear - Ursa Major, and Arcas is the Little Bear - Ursa Minor.

The stars in the bowl of the Little Dipper provide an easy way to gauge star visibility when you are out viewing. The four stars in each of the four corners of the bowl range in magnitude from 2 to 5, they form a convenient scale: Kochab - 2.1, Pherkad - 3.1, Zeta - 4.3, and Eta - 5.0. If you can see all four stars in the bowl, you are seeing stars down to 5$^{th}$ magnitude.

The bowl of the Little Dipper crosses the meridian just after dark around June 25th. The meridian is an imaginary line running from north to south across the sky and passing through the zenith. At this time the stars within this asterism are in their highest position above the horizon.

Contrary to popular belief, **Polaris** is not the brightest star in the northern sky. At magnitude 2.0, it is just the 49th brightest star in the sky, (2.0; white supergiant). See the *Helpful Charts and Data* chapter for a list of the 10 brightest stars.

Near Polaris is a small ring of 7th & 8th magnitude stars known as the "engagement ring", with Polaris in the place of the diamond. This ring is visible with binoculars on a clear, dark night. See if you can find it.

**Kochab** is the brightest star in the bowl of the Little Dipper, (2.1; orange giant).

**Pherkad**, Arabic for "two calves", is the second brightest star in the bowl, (3.1; blue-white giant).

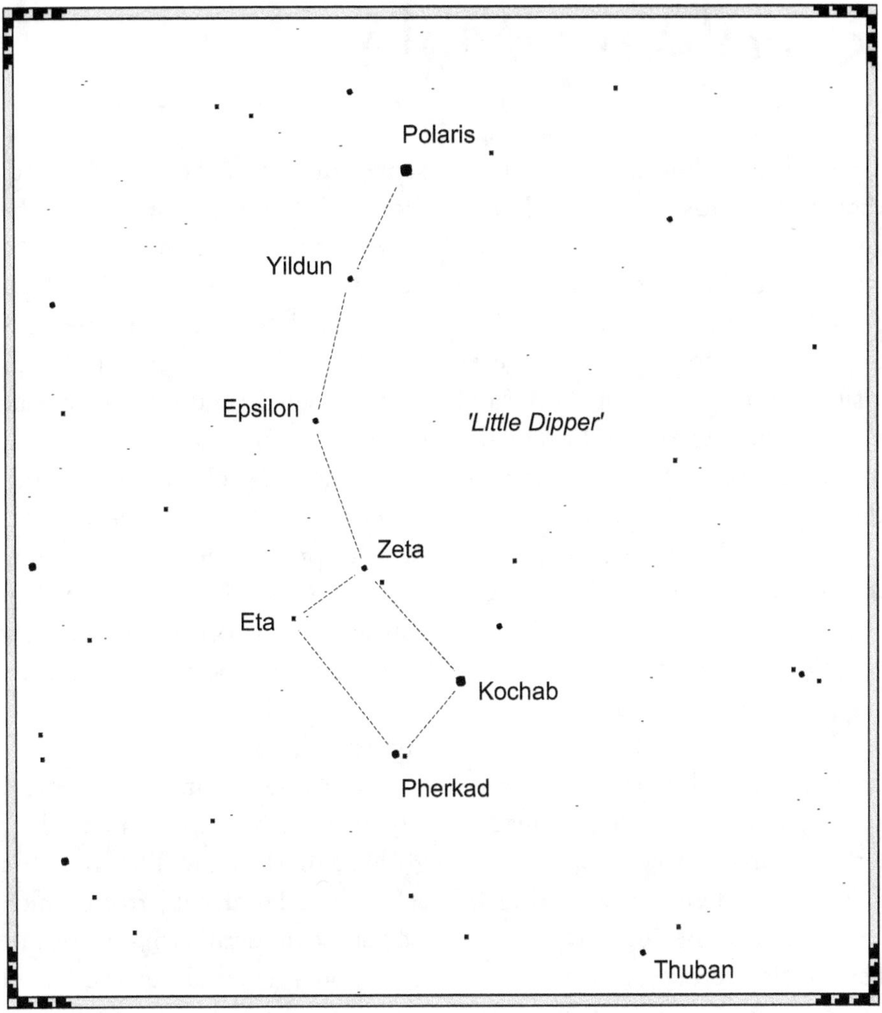

Polaris

Yildun

Epsilon

*'Little Dipper'*

Zeta

Eta

Kochab

Pherkad

Thuban

Detail view of Ursa Minor

**Zeta UMi**, a blue-white star, is the third brightest star in the bowl, (4.3; blue-white star).

**Eta UMi** forms the fourth corner of the bowl and is the dimmest of the four bowl stars, (5.0; white star).

The two stars in the middle of the handle are **Epsilon UMi**, (4.3; blue-white giant), and **Yildun**, from the Turkish word for "star", (4.4; blue-white star).

**Thuban** is part of the constellation Draco, the Dragon (3.7; blue-white giant) .

# Revolving Male

## Náhookǫs biką'ii

Revolving Male is a star figure that circles closely around North Fire. A similar figure called Revolving Female lies on the opposite side of North Fire. Together they form a pair of figures visible from the Navajo homeland year round. They symbolize the importance of a family created by the union of a husband and wife.

Revolving Male's circular path of motion is similar to the circular shape of a traditional Navajo dwelling, the  hooghan. These celestial references to the hooghan and husband and wife serve to reinforce the importance of family to the Navajo people.

The complementary pairing of Revolving Male and Revolving Female also represents the law that two couples should not live together or share the same cooking fire.

Many American Indians saw this group of stars as a bear and compared their rising and falling motions to that of a real bear's hibernation cycles. The Micmac people of Nova Scotia described this cycle as a never-ending bear hunt (see Other Stories chapter). To them the rising of the bear in spring represents a reawakening or rebirth from hibernation. Summer is the time of the chase. In the fall, the bear is mortally wounded; it's blood colors the leaves of the maple trees red. Winter represents the bear's death. Each year the cycle repeats itself.

The Sioux people in central North America saw these stars as a skunk, with it's prominent tail held high.

The Cherokee people, in southeastern North America, saw these stars as a bear (the four stars in the bowl) pursued by three hunters (the handle stars).

The Aztec people of Central America knew these stars as their god Tezcatlipoca. The Mayans called him Hunracan, from which we get the modern term - hurricane. This god rushed about the world causing trouble until his good brother tricked him and placed him in the sky to dance forever, where he couldn't cause more trouble in the world.

Revolving Male is essentially the asterism known as the Big Dipper today. It is one of the most widely recognized star patterns in the northern hemisphere. The "dipper" reference may have originated with

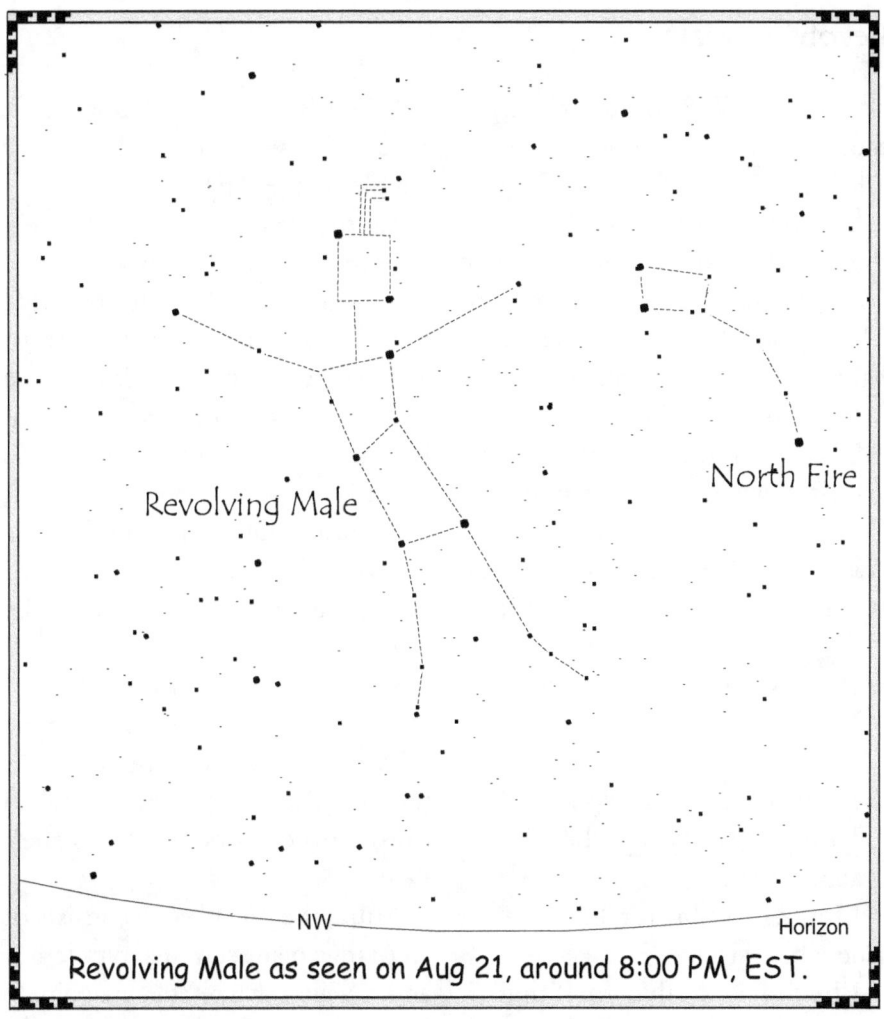

Revolving Male as seen on Aug 21, around 8:00 PM, EST.

## Finding Revolving Male:

Revolving Male is circumpolar for viewers north of 45 degrees latitude. That means it circles the northern celestial pole and never drops below the true horizon. An obscured true horizon may make it hard to see during November and December.

| WHERE TO LOOK 1 HOUR AFTER SUNSET | | | |
|---|---|---|---|
| WINTER | SPRING | SUMMER | FALL |
| Low in N, | High in NE, | High in NW, | Setting in NW |
| Dec 21 | Mar 21 | Jun 21 | Sep 21 |

## Revolving Male

slaves in the U.S. fleeing north on the Underground Railroad in the mid 1800's.

The British know these stars as the Plough. Other European people know them as the Great Wagon and Three Horses. The middle star in the handle and its companion are known as Jack on the Middle Horse.

The Big Dipper is actually an asterism within the constellation Ursa Major, the Great Bear. Ancient Arab astronomers also saw the figure of a bear in this constellation and named many of its stars according to parts of a bear's anatomy.

Four stars make up the bowl of the Big Dipper:

**Dubhe** is Arabic for "bear", (1.8; orange giant).

**Merak** is Arabic for "flank", (2.4; blue-white star).

**Phecda** is Arabic for "thigh", (2.4; blue-white star).

**Megrez** is Arabic for "root of the tail", (3.3; blue-white star).

Dubhe and Merak are known as the Pointer Stars. An imaginary line drawn from Merak through Dubhe points to Polaris, the North Star.

**Alioth** is Arabic for "bull", (1.8; blue-white star). It lies in the handle of the dipper.

**Mizar** is Arabic for "groin", (2.3; blue-white star). It also lies in the handle of the dipper.

**Alcor,** "the overlooked one", appears to be a close companion to Mizar but is actually unrelated, (4.0; blue-white star). Both the Arabs and some Native Americans used Alcor as a test of eyesight. This pair is sometimes called the "Horse and Rider".

**Alkaid,** Arabic for "chief of the mourners", (1.9; blue star). Alkaid lies at the end of the handle of the Big Dipper.

Below the bowl of the Big Dipper there are three pairs of stars forming a line  roughly the same length as the Big Dipper. The Arabs called these three pairs the "Gazelle's Leaps".

**Alula,** the "first leap", consists of an orange giant star, magnitude 3.5, and a yellow-white star, magnitude 3.8.

**Tania,** the "second leap", consists of a blue-white subgiant star, magnitude 3.5, and an orange-red giant star, magnitude 3.1.

**Talitha,** the "third leap", consists of a blue star of magnitude 1.9, and a blue-white star, magnitude 3.6.

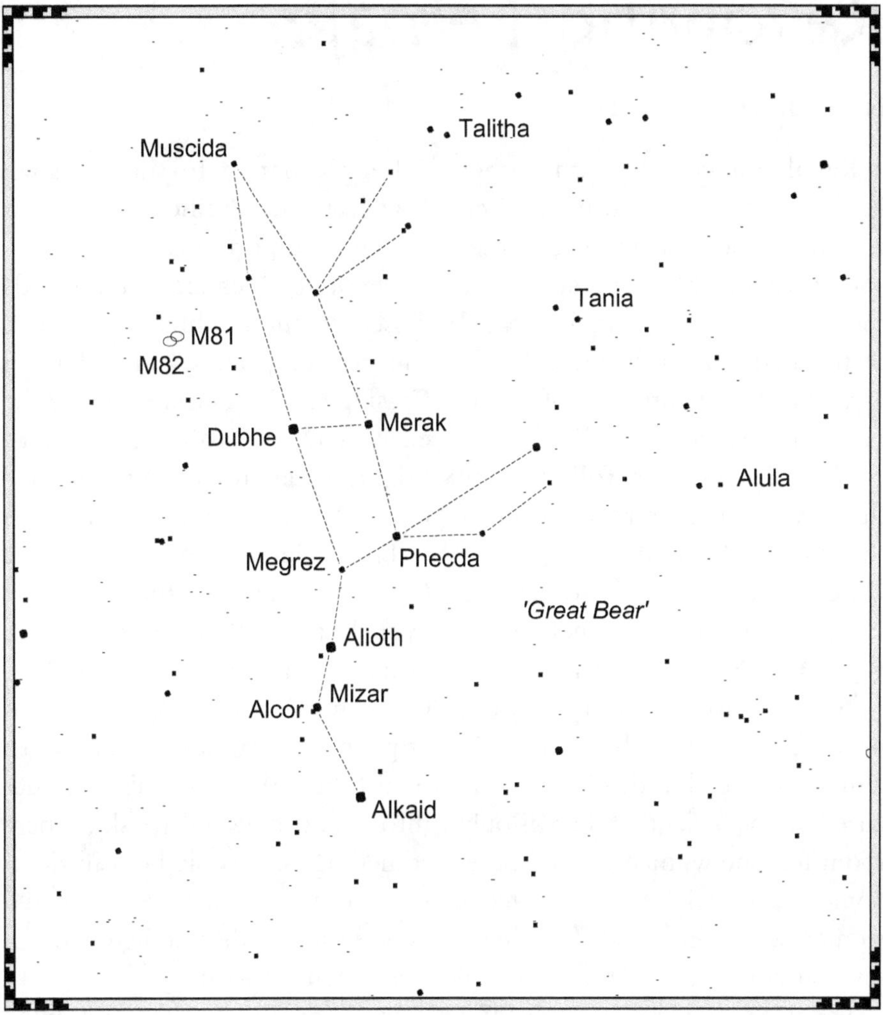

Detail view of Ursa Major

**Muscida,** Latin for "muzzle", forms the nose of the bear in the constellation Ursa Major, (3.4; yellow-white giant).

**M81** is a spiral galaxy of 7th magnitude. On clear, dark nights you might be able to see it as a faint smudge located on a line running through Phecda and Dubhe. It is the third brightest galaxy in the sky, often visible with binoculars. **M82** is a smaller galaxy.

# Revolving Female

## Náhookǫs ba'áadii

Revolving Female is the other half of a pair of figures circling closely around North Fire. Together with Revolving Male they symbolize the importance of a family created by the union of a husband and wife, as well as the significance of family duties and responsibilities.

The circular path followed by these two figures is reminiscent of the circular shape of a hooghan.

Revolving Female is sometimes  seen as a reminder of the Navajo custom that a mother-in-law should never be in a hooghan with her daughter's husband. This law is based on the belief that there is only room for one woman in the home. Some traditional Navajo believed a woman would go blind if she ever saw her son-in-law.

The five stars in a distinctive "W" shape were seen by the Lapps of northern Europe as part of a moose's antler. Siberia's Chukchee people saw them as a group of five reindeer.

Some Arab people in the Middle East pictured them as the humps on a camel's back.

People native to the Marshall Islands in the Pacific saw them as the tail of a porpoise.

Revolving Female corresponds closely to the brighter stars within the contemporary constellation Cassiopeia, the Queen. During the early winter months, the five brightest stars resemble a squashed letter "M" in the northern sky.

The Greek legends describe Cassiopeia as a boastful queen. She even boasted that she was more beautiful than the gods themselves. Of course the gods did not appreciate this and decided to punish her. She was chained to her throne and hung in the sky, where sometimes she would be seen right-side up and at other times she would be seen in a rather undignified up-side down position.

In China, these stars represented Wang Liang, a famous charioteer in the 11th century B.C., and his chariot.

**Caph**, Arabic for "hand", (2.3; white subgiant).

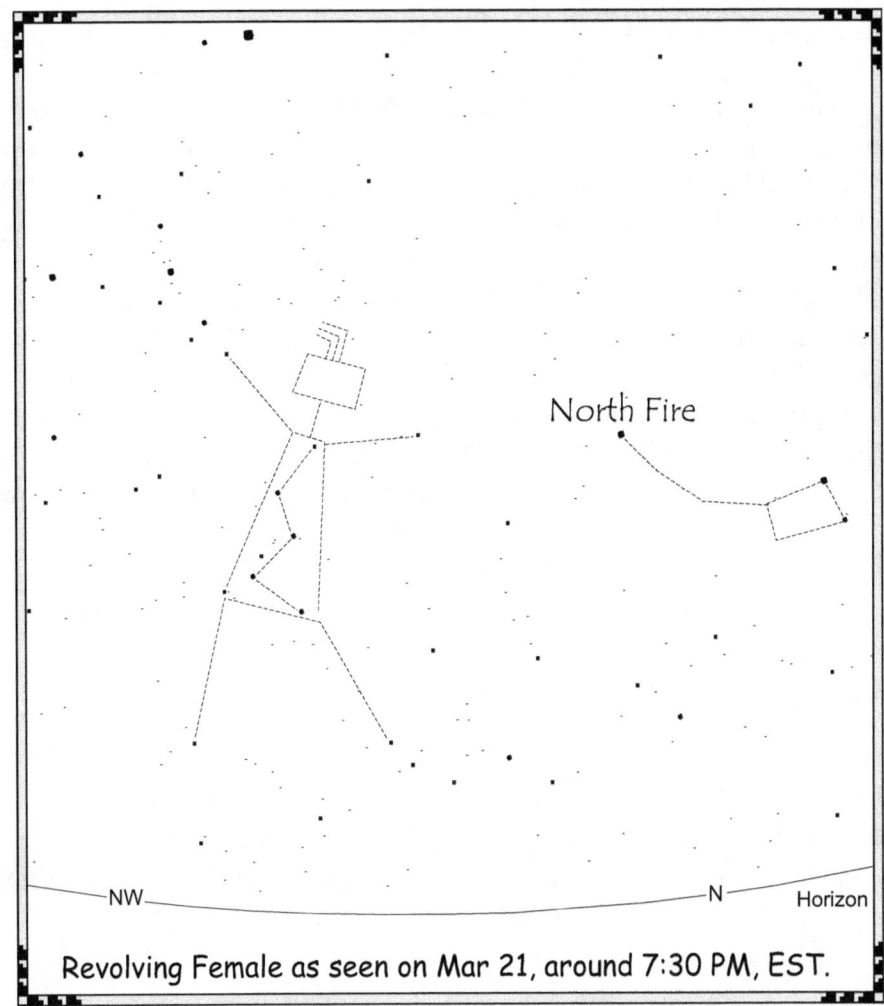

North Fire

**Revolving Female as seen on Mar 21, around 7:30 PM, EST.**

### Finding Revolving Female:

For locations north of 50 degrees latitude, Revolving Female is always visible above the true horizon. Depending on how clear your view of the true horizon in the north is, you may have trouble seeing Revolving Female during the summer months.

## WHERE TO LOOK 1 HOUR AFTER SUNSET

| WINTER | SPRING | SUMMER | FALL |
|---|---|---|---|
| Overhead, | High in NW, | Low in N, | High in NE, |
| Dec 21 | Mar 21 | Jun 21 | Sep 21 |

## Revolving Female

**Schedar**, Arabic for "beast", (2.2; orange giant).

**Cih**, Chinese for "whip", is an unpredictably variable star that averages a magnitude of 2.5, (blue subgiant).

**Ruchbah**, Arabic for "knee", (2.7; blue-white giant).

**Segin** lies at the flattened end of the "M" shaped group of stars, (3.4; blue giant).

Revolving Female lies within the bright swath of stars we know as the Milky Way. Looking at this area with binoculars will reveal thousands of dim stars that appear to the naked eye as a bright haze, as if there were a thin cloud across the sky.

**M31**, the Great Andromeda Galaxy, actually lies within the constellation Andromeda, the Princess. If you follow a line in the direction where Cih, Schedar, and Caph seem to be pointing, you may find a faint, hazy oval object. This spiral galaxy, similar to our own, is the most distant object visible with the naked eye. It lies about two million light years away. Binoculars can enhance the view, maybe even allowing you to see the 9th magnitude galaxy **M32**, which lies next to M31.

The house-shaped asterism within the constellation **Cepheus**, the King, lies nearby. According to Greek legend, Andromeda was his daughter, the princess, and Cassiopeia, the Queen, was his wife.

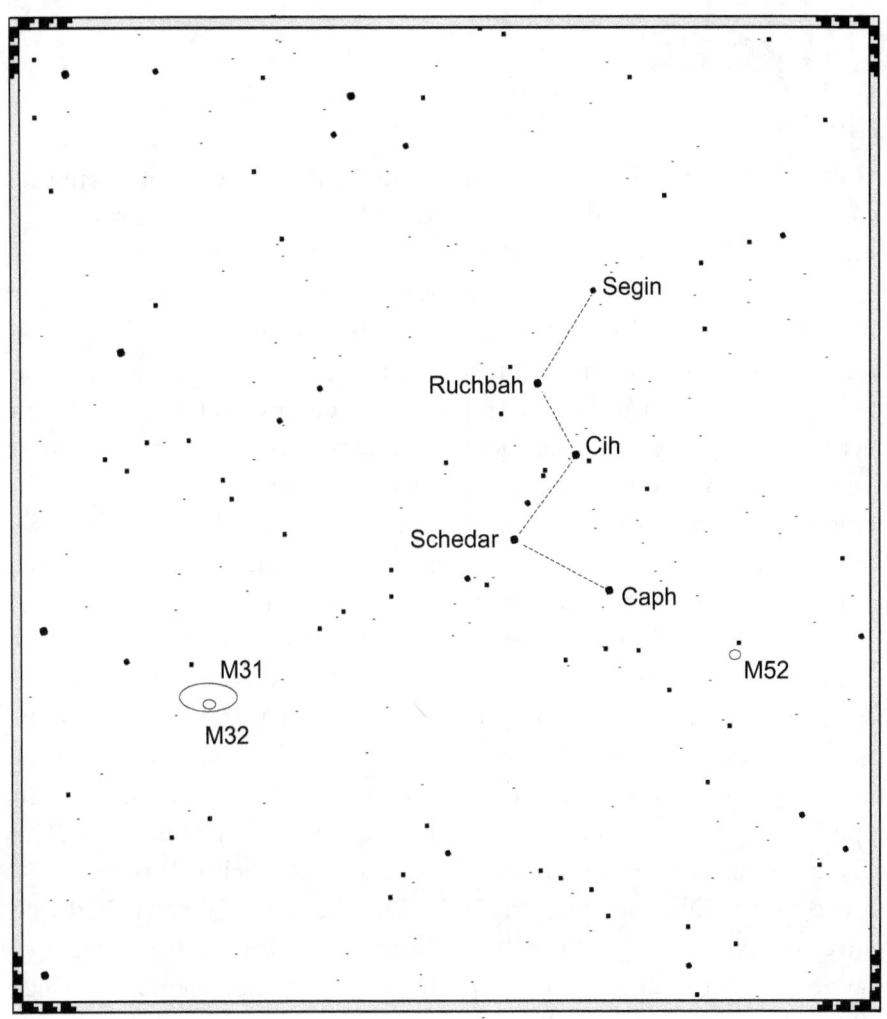

Detail view of Cassiopeia

# Dilyehe

## Dilyéhé

Dilyehe was the first group of stars created and placed by Black God. Black God unveiled "stars" to the other Holy People for the first time when he arrived with Dilyehe on his ankle. The sparkling light in the stars was a sign of his control over fire and light. By making these stars jump to his forehead, he demonstrated his power to place them in the night sky.

The other Holy People agreed that Black God's placement of Dilyehe proved his ability to beautify the night sky, so they had him place the stars in the "dark upper".

There is no common translation of the name Dilyehe, but some sources describe it as "pin-like sparkles" or "sparkling figure".

The Navajo used Dilyehe's appearance as a temporal marker. When these stars were first seen rising at twilight, around mid-October, it marked the beginning of the season when certain sacred ceremonies could be performed. Healers and chanters noted Dilyehe's position in the sky during the night to mark the passage of time so that ceremonial activities would end precisely at dawn.

The end of this ceremonial season occurs in March, when the first thunderstorms appear. Dilyehe is seen setting at twilight at this time.

This ceremonial season closely corresponds with the Navajo's "winter" season. Traditionally, the Navajo divded the year into two seasons: summer, *Hai*, and winter, *Shi*, reflecting the principle of complementary pairs.

Dilyehe is also associated with the time of year that we know as the month of July. At this time, Dilyehe appears in the sky just before the light of dawn hides the stars. This is known as a heliacal rising.

Dilyehe is sometimes paired with First Slim One to form the two halves of a complementary pair that are related to agricultural signs. When First Slim One is seen setting at twilight, in early May, the planting season begins. When Dilyehe is seen rising just before dawn in late June, the time for planting is over.

Some Native American stories describe these stars as children who wanted to walk among the stars. They managed to find a way to the

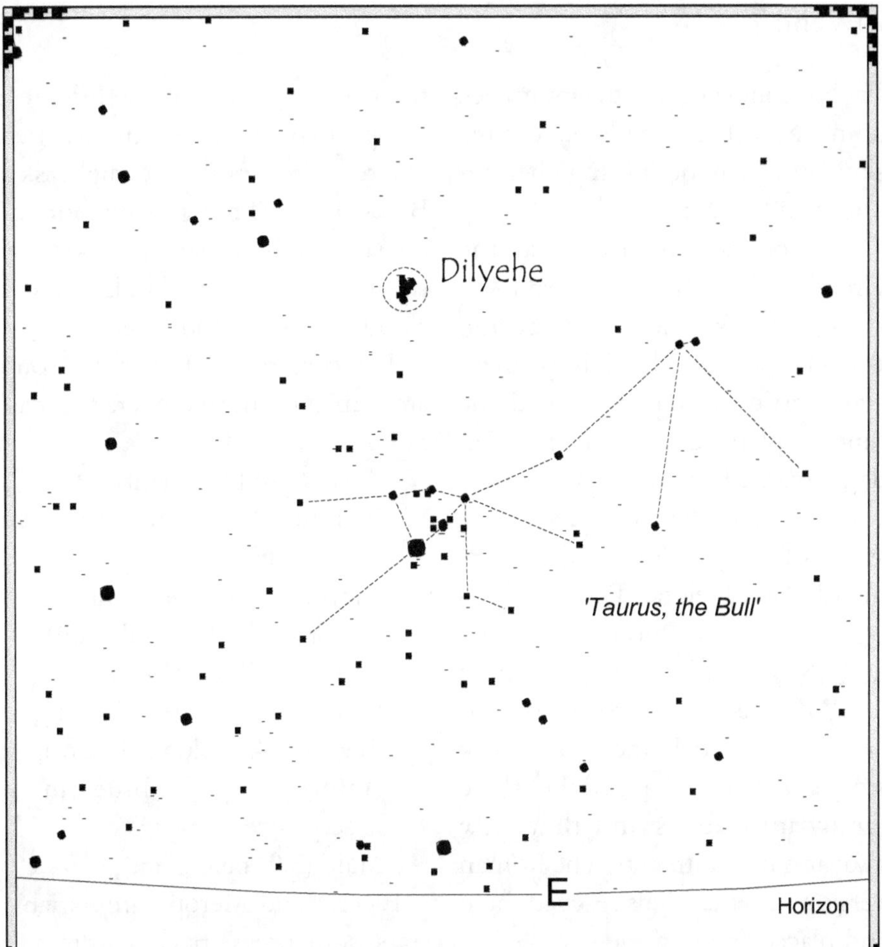

Dilyehe

'Taurus, the Bull'

E

Horizon

Dilyehe as seen on Dec 21, around 6:00 PM, EST.

| WHERE TO LOOK 1 HOUR AFTER SUNSET | | | |
|---|---|---|---|
| WINTER | SPRING | SUMMER | FALL |
| High in E, Dec 21 | High in W, Mar 21 | Setting WNW May 10 | Rising ENE, |

## Dilyehe

sky, but could not find their way back home. Now they wander across the sky forever, huddling together for safety and warmth.

It is curious to note that the Greeks and Romans also used these stars to mark when to plant and when to harvest. The sight of these stars setting at dawn marked the time to plant. Seeing them rise at dawn marked the time to harvest.

Greek mythology describes these stars as the Seven Sisters, daughters of Atlas and Pleione. The story tells  how Orion grew overly fond of the daughters and tried to break into their home. The Greek story says Venus changed the whole family into doves and they flew away to safety inthe sky. The Roman version says that Zeus rescued them and placed them in the sky.

To astronomers, this cluster of stars is known as **M45**, the Pleiades. It lies in the consteallation Taurus, and is made up of hundreds of stars, but only seven or so are visible to the naked eye. This is the brightest and probably the most famous open cluster of stars in the night sky. Binoculars will bring many more of the stars in this cluster, and possibly some of the faint nebula, or gas cloud, that surrounds them.

The brightest stars in this group are named for the members of Atlas' family:

Atlas (3.6, blue giant),
Pleione, Atlas' wife (4.8, blue subgiant),
Alcyone (2.9, blue giant),
Merope (4.2, blue subgiant),
Electra (3.7, blue giant),
Celaeno (5.5, blue subgiant),
Taygeta (4.3, blue subgiant),
Asterope (5.8, blue main sequence star),
Maia (3.9, blue giant),

Electra and Merope are variable stars, sometimes appearing dimmer than at other times. The myths attribute this dimming to their occasional crying, which blurs the brilliance of their eyes.

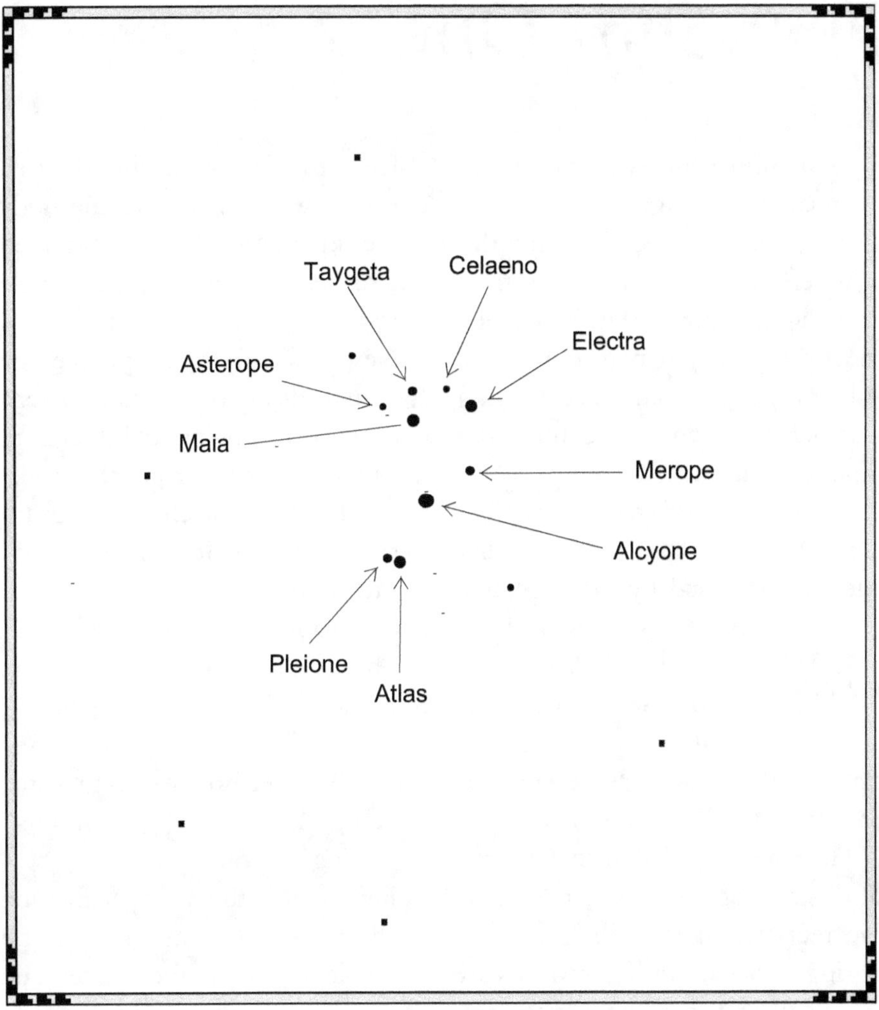

Taygeta    Celaeno

Electra

Asterope

Maia

Merope

Alcyone

Pleione

Atlas

Detail view of Pleiades

# First Slim One

## 'Átsé'ets'ózí

First Slim One is a star figure associated with agriculture. He is also known as the keeper of months.

When this figure was seen setting at twilight, in early May, it signaled the time to begin planting crops. When nearby Dilyehe is seen rising just before dawn in late June, the time for planting is over.

A curved line of stars beside First Slim One is often seen as a digging stick, a tool used by many Native  Americans to prepare the ground for planting seeds. See if you can spot a small group of stars nearby that might resemble a basket of seeds.

The reference to 'slim' in the name of this star figure may be related to the fact that it is visible at night during the lean winter months of the year, when most plants are dormant and food for the people is scarce.

First Slim One is often associated with the star figure First Big One, forming the two halves of a complementary pair. First Big One is located on the opposite side of the sky and is visible at night during the summer months.

In central Brazil, the Bororo people saw these stars as the body of a cayman, one of the most feared and respected animals in their jungle world.

New Zealand's Maori people saw the three distinct stars as the stern of a canoe belonging to Tamarereti, a fisherman who broke the gods' rules of fishing. He choked to death eating the forbidden fish. His position in the sky reminds the people to respect these rules.

This is the same group of stars that the Arabs called "Al Jabbar", "the giant" or "the great". They called the three bright stars in a line, forming his belt, the "golden nuts" or "string of pearls".

The stars within First Slim One are better known as part of the constellation Orion, the Hunter, one of the more recognizable constellations due to it large size and unmistakable line of three bright stars. In Greek mythology, Orion became so boastful of his hunting skills that the gods sent a tiny scorpion to kill him. Orion succeeded in killing the scorpion, but was fatally stung in the process. The gods placed them both in the

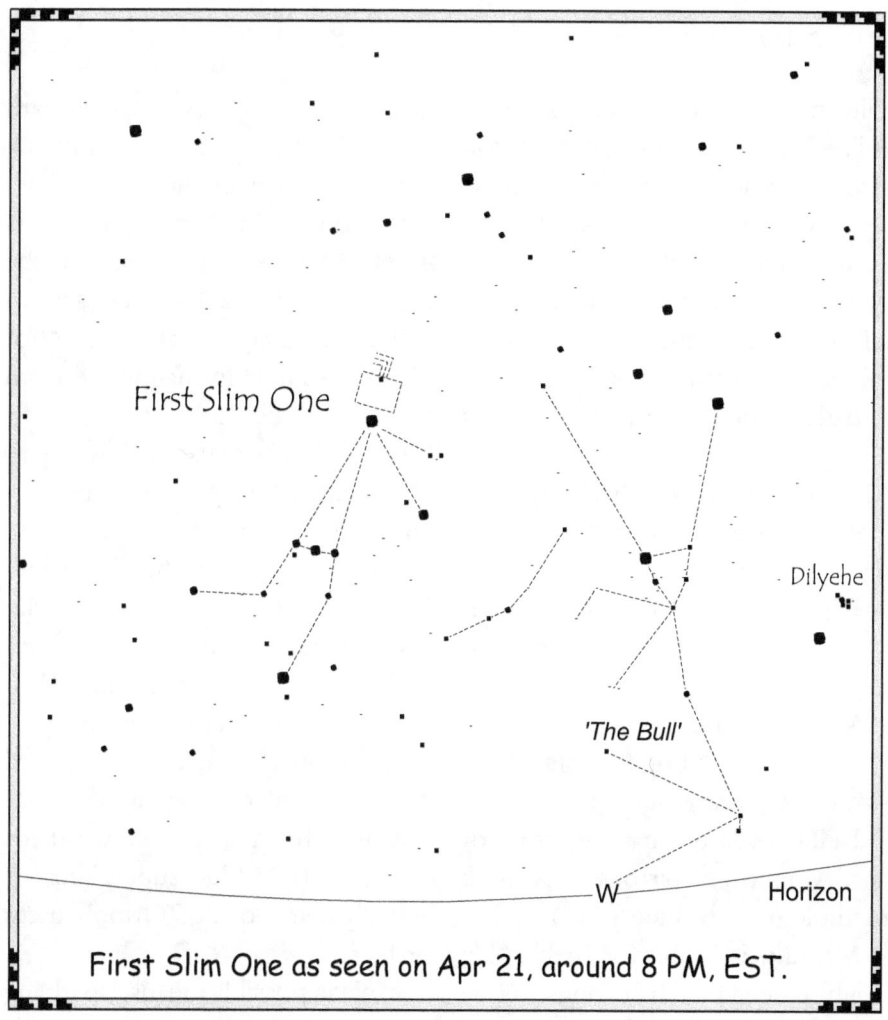

**First Slim One as seen on Apr 21, around 8 PM, EST.**

heavens but on opposite sides of the sky so that they would never be seen at the same time. The stars in the constellation Scorpius are known to the Navajo as First Big One.

This particular group of stars comes closer than most to actually resembling a person. One star marks his beard, two more mark his shoulders, and three bright stars in

| WHERE TO LOOK 1 HOUR AFTER SUNSET | | | |
|---|---|---|---|
| WINTER | SPRING | SUMMER | FALL |
| Rising E, | Setting W, | Below NW | Below NE |
| Jan 1 | May 1 | horizon | horizon |

## First Slim One

a line mark the belt around his waist. A line of three fainter stars mark the sword or scabbard hanging from his belt. Two other stars mark his knees or feet. To his right, a curved line of stars represents his shield. To the left, more stars mark his other arm, raised overhead as if wielding a club.

**Betelgeuse**, (BET-el-jooze), is Arabic for "house of the twins", (0.5; orange-red bright giant). It's diameter is approximately the size of Venus' orbit!

 **Rigel** is derived from a 10th century Arabic word for "foot", (0.1; blue-white supergiant). It is the sixth brightest star in the sky, not counting our sun.

**Bellatrix** is the name of one of the Amazon warriors in Greek mythology, (1.6; blue giant).

**Mintaka** is Arabic for "belt", (2.2; bright blue giant). It belongs to one of the hottest types of stars, ranging up to 45,000 degrees Kelvin. By comparison, our sun is only 6,000 degrees Kelvin.

**Alnilam** is Arabic for "string of pearls", (1.7; blue supergiant).

**Alnitak** is derived from another Arabic word for "belt", (1.8; blue supergiant). It is even larger and hotter than Mintaka. It has a 4th magnitude companion that is hard to see due to Alnitak's brightness.

In the center of the "sword" are two "stars" known together as **Theta Orionis**. The "star" on the right is actually a four-star system called the Trapezium. These hot blue stars range in magnitude from 5 to 7 and lie 1500 ly away. The "star" on the left is actually a pair of 5th and 6th magnitude blue stars, about 2600 ly away.

All of the stars in the Trapezium lie within the Orion Nebula, **M42**, a vast greenish-white cloud of gas and dust visible even to the naked eye. They are among the youngest stars in our galaxy, having formed just a few million years ago.

**Saiph**, from an Arabic word for "sword", (2.1; blue supergiant). It actually marks one of Orion's knees or feet.

**Meissa** possibly means "sparkling one", (3.4; blue star).

**Sirius**, the Dog Star, lies within Canis Major, the Great Dog, one of two dogs who were Orion's hunting companions. It is the brightest star in the night sky, and the fifth closest at 8.6 light years away, (-1.46; blue-white star). It is located to the left of the Orion figure.

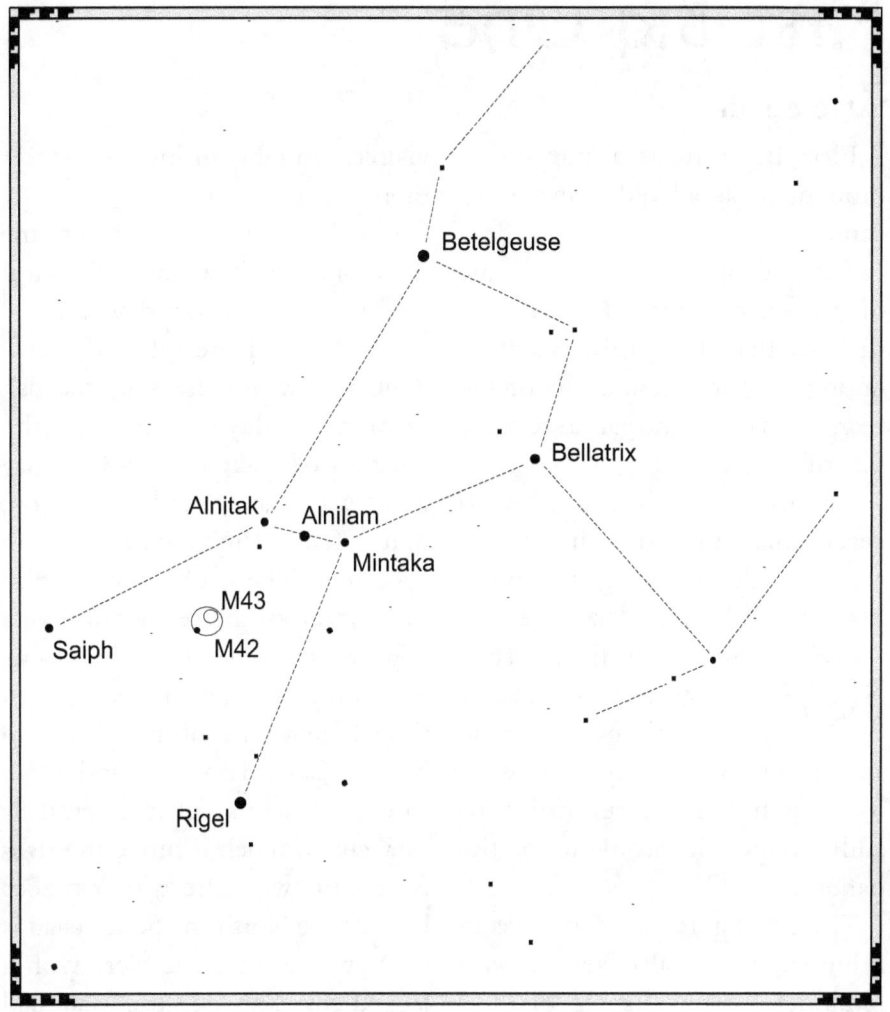

Detail view of Orion

# First Big One

## 'Átsé'etsoh

First Big One is a star figure thought of as an old man with a cane.

The reference to 'big' in the name of this star figure may be related to the fact that it is visible at night during the summer months of the year, when crops and game are more plentiful.

According to some Navajo ceremonial traditions, individual stars in the star figures were identified as specific parts of the figures, such as legs, head, organs, or even it's cane. These associations with body parts were used in healing ceremonies to address specific problems of the patient.

The star figures' resemblance to a human figure also served as a reminder for the Navajo of their close relationship to the Holy People who created the star figures and to the world in which they lived.

First Big One is often associated with First Slim One, forming two halves of a complementary pair. First Slim One is a star figure on the opposite side of the sky and is visible at night during the winter months.

First Big One is also linked to the the month of December, when it it seen rising just before dawn.

The Maori people of New Zealand saw in these stars the fish hook that played a part in the creation of their islands. One day Maui was fishing with this hook, made from the jawbone of an ancestress. The hook got caught on the bottom of the sea. Maui gave a mighty tug on the line and up came a giant fish, as big as an island. Even though he warned his people not to harm it, they did not listen and began to cut off chunks of it. Eventually they cut so much it broke the fish-island in two. These two pieces became the North and South Islands of New Zealand, complete with a jagged coastline. Because Maui had pulled so hard on the line, when the hook broke free, it flew all the way up into the sky where it can still be seen today.

Many people in the Indonesion archipelago saw a palm tree in the upper portion of this group of stars.

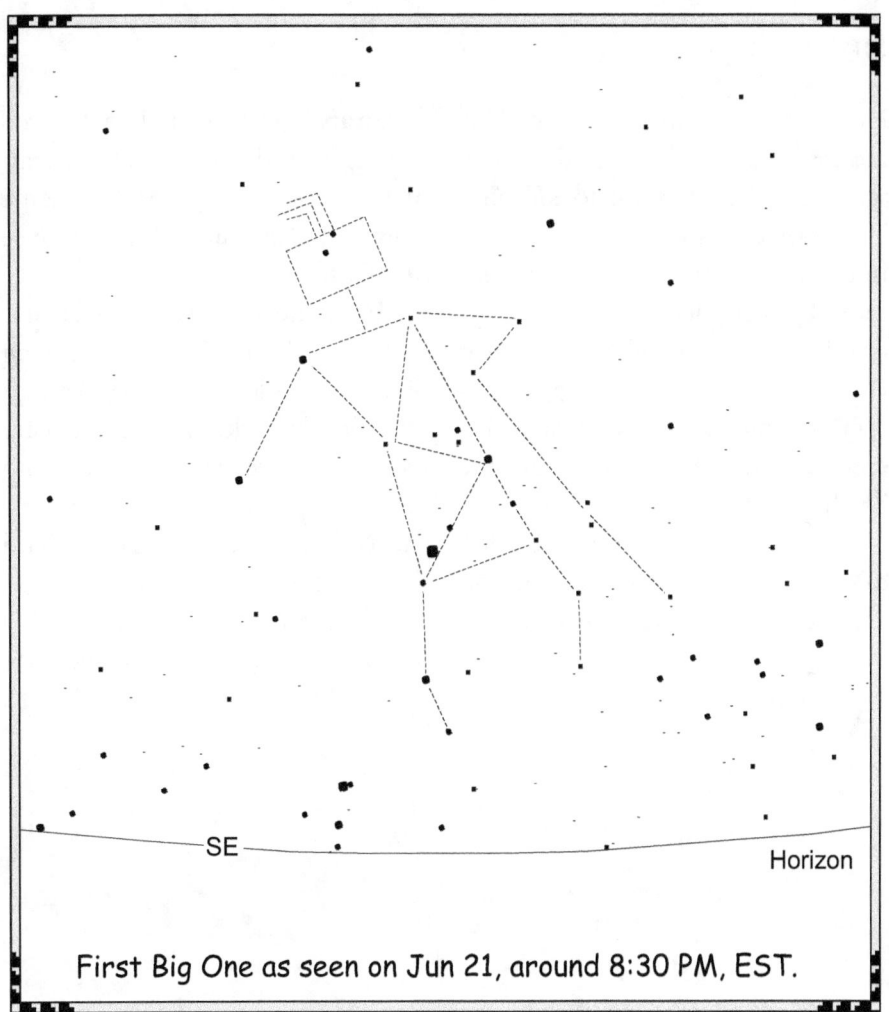

**First Big One as seen on Jun 21, around 8:30 PM, EST.**

The stars in the lower part of First Big One are part of the Greek constellation Scorpius. This constellation is named after the scorpion sent to kill Orion because he boasted of his hunting skills too much. Orion even claimed he was such a great hunter that he would

| WHERE TO LOOK 1 HOUR AFTER SUNSET | | | |
|---|---|---|---|
| WINTER | SPRING | SUMMER | FALL |
| Below W horizon | Below E horizon | Rising SE, Jun 21 | Setting SW, Sep 21 |

## First Big One

kill all of the animals on earth. This angered Gaia, goddess of the earth, who sent the scorpion to kill him. The scorpion stung Orion on the heel and he fell to the ground mortally wounded.

When the constellation Orion (First Slim One to the Navajo) is seen setting, the stars of Scorpius are seen rising on the opposite side of the sky.

The four stars that form the end of the upturned tail of the scorpion are known by the Navajo as the star figure Rabbit Tracks.

**Antares** (variable 0.9 to 1.8, red supergiant)is the red star at the heart of the scorpion figure. The name Antares means "like Mars" because of this star's red color.

**M4** is the nearest globular star cluster to Earth. This hazy looking object is made up of thousandsof stars closely packed in a tight cluster formation. Because it is about 6th magnitude, and the stars are loosely packed in the cluster, it is often hard to spot, even with binoculars.

**M7** is an open cluster of 3rd magnitude. This cluster of stars can normally be seen in detail with binoculars.

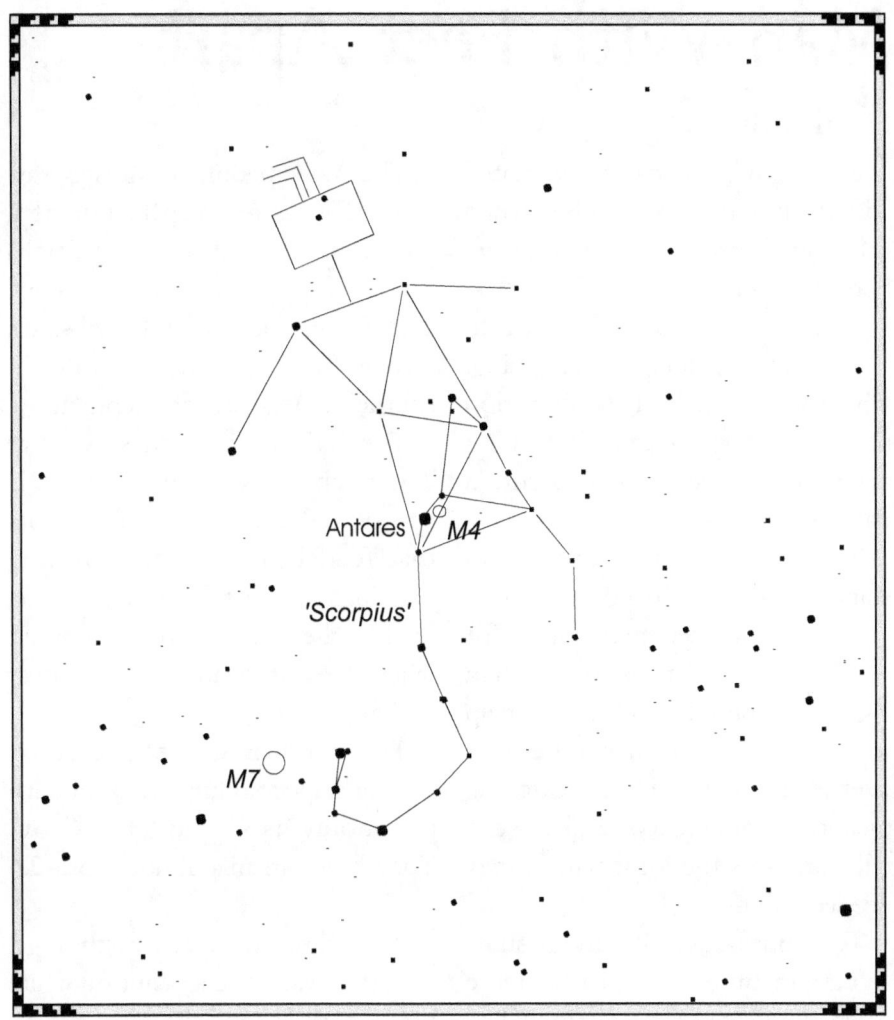

Detail view of Scorpius

# Man With Feet Apart

## Hastiin Sik'ai'í

Man With Feet Apart is associated with the month of November, when this star figure is seen rising just before dawn.

Feet Apart is also associated with the Navajo practice of stargazing, which was a method of divination used to determine which healing ceremonies were needed to cure a patient's illnesses.

The star figure is a human form standing in a braced position, with  his feet spread apart for stability, and his arm raised to his brow to help him see farther. Because Feet Apart is in such a searching position, he is able to see all and can tell stargazers the information they needed to know.

The appearance of certain stars, at certain times of the year were often interpreted as predictions. The stars of Feet Apart were sometimes use to predict the quality of the fall harvest. If the stars were clear and bright, the harvest would be good. But if the stars appeared dim and fuzzy, the harvest and the weather at harvest time would not be very good.

The Aztec people of Mexico saw these four stars as the tail of a scorpion. The Wailwum of Australia saw a kangaroo in the these stars. The Tukano people of Brazil saw a heron. The Khmer people in southeast Asia saw an elephant.

The Chinese knew these stars as T'ien-tche, the celestial cart. It epresented the carts and chariots that feudal princes used to bring offerings to Great Feast of Pleasure, held in the last month of summer, when these stars are seen setting at twilight.

The four main stars of Feet Apart form a trapezoid known in modern astronomy as Corvus, the Crow. They range in magnitude from 2.6 to 4.0.

According to Greek mythology, the crow was once a beautiful white bird with a lovely voice. He belonged to Apollo, who one day sent him to fetch a drink of water. The crow became distracted on the way by a fig tree and waited until the figs were ripe.

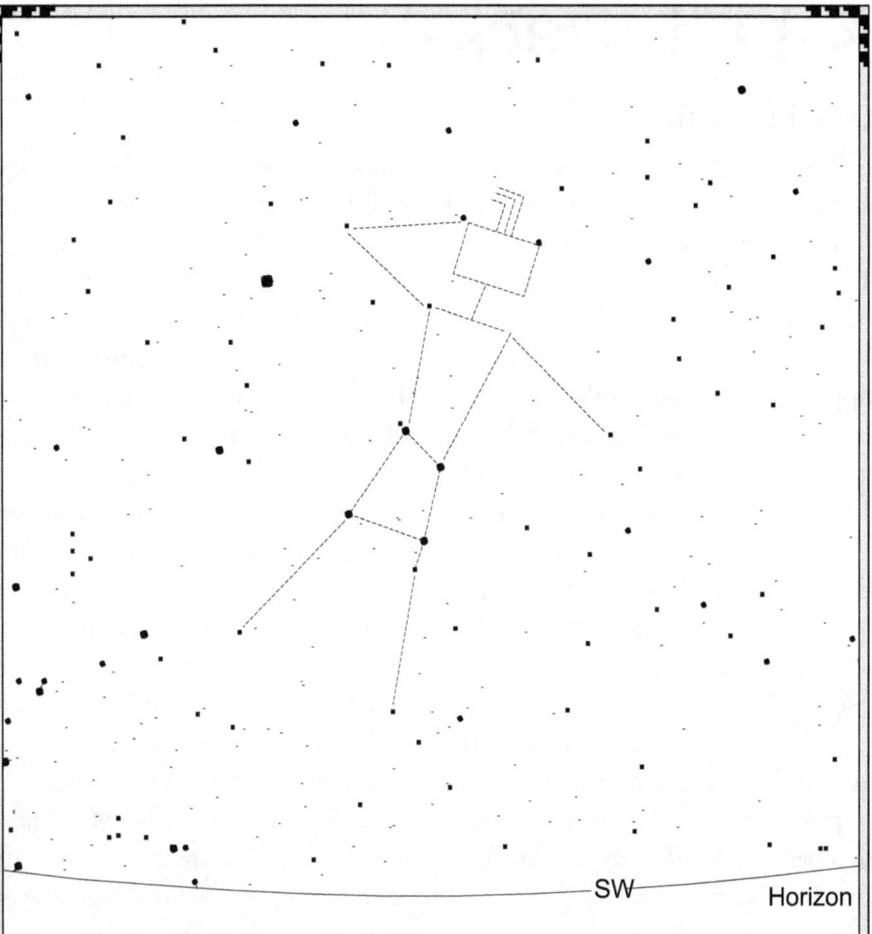

Feet Apart as seen on Jun 21, around 8:30 PM, EST.

By the time the crow returned, Apollo was so angry that he turned the crow black and changed his voice to a screech. He then set the crow in the sky, close to but never reaching the constellation Crater, the Beaker of Water.

| WHERE TO LOOK 1 HOUR AFTER SUNSET | | | |
|---|---|---|---|
| WINTER | SPRING | SUMMER | FALL |
| Below N horizon | Rising ESE, Apr 1 | High in S, May 21 | Setting WSW Aug 1 |

# Rabbit Tracks

### Gah heet'e'ii

Rabbit Tracks is a set of four stars, arranged in a pattern resembling a set of rabbit tracks. This star figure is associated with hunting.

When these stars were seen tipping to the east, in the early fall, it signaled when Navajo hunters could start pursuing the deer and antelope. The young animals of these species are no longer dependent on their mothers at this point and can look after themselves.

By following this guideline the Navajo ensured that deer and antelope populations would be able to sustain themselves, and in turn provide food for the Navajo.

Rabbit Tracks also honors the rabbit, which the Navajo depended on as one of their primary sources of food. Placing the rabbit's tracks in the sky kept them in a visible place that would constantly remind the people of their dependent relationship with the world around them.

The people of Java saw these stars as a goose that nests under a plam tree, which is represented by the upper part of the "scorpion".

In modern astronomy, these stars are known as the tail of the scorpion in the constellation Scorpius.

Nearby is an open cluster of about 50 stars known as **M7**. These are about 3rd magnitude and with good viewing conditions can be seen with the naked eye as a fuzzy-looking star.

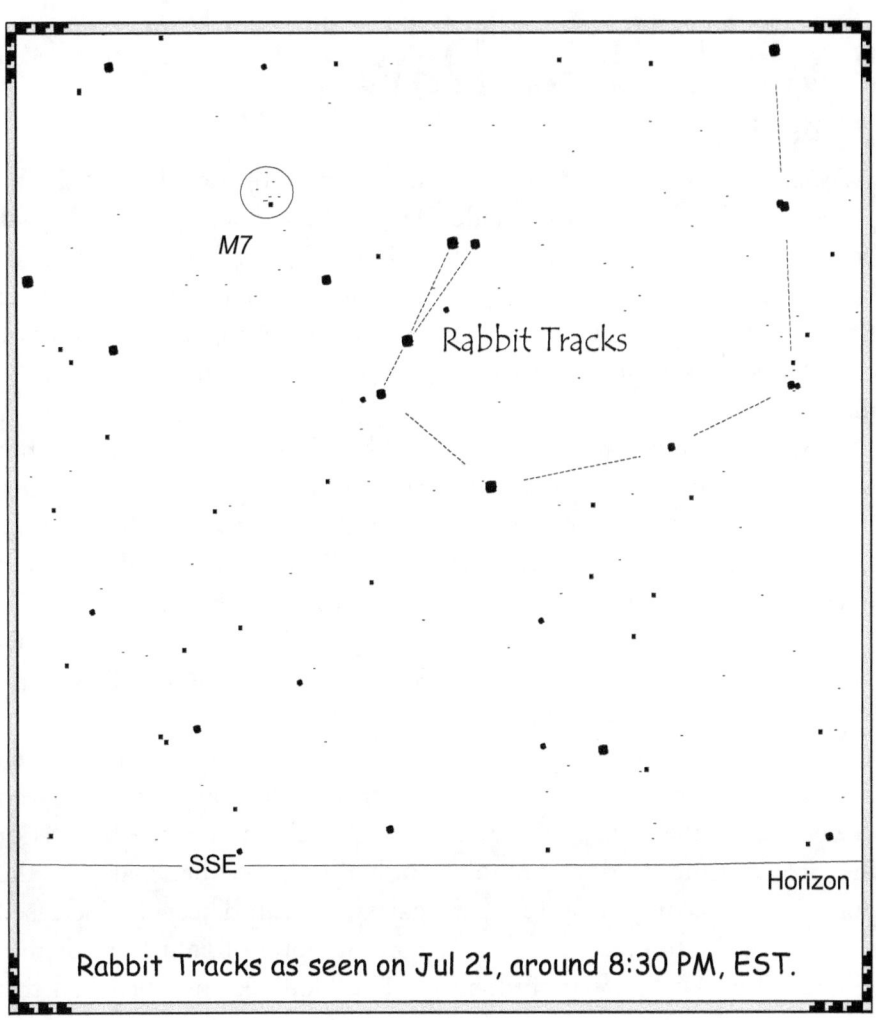

M7

Rabbit Tracks

SSE

Horizon

Rabbit Tracks as seen on Jul 21, around 8:30 PM, EST.

| WHERE TO LOOK 1 HOUR AFTER SUNSET | | | |
|---|---|---|---|
| WINTER | SPRING | SUMMER | FALL |
| Setting SW, | Below N | Rising SE, | Low in S, |
| Nov 1 | horizon | Jun 21 | Aug 15 |

# Awaits the Dawn

## Yikáísdáhí

Awaits the Dawn is the last star placement that Black God made before stopping to rest. Instead of being a small group of stars, this star figure is actually a scattering of countless small stars in a path across the sky.

The name originates from the way that Awaits the Dawn rises before dawn along the eastern horizon in January. Because this star figure resembles a bright ribbon of stars across the sky, it appears as a false dawn. The white light of dawn is one of the four cardinal light phenomena recognized by the Navajo. Dawn is the proper time of day for holy prayers and making plans for that day and for the future.

The blue sky light of midday represents the proper time for productive activity. The yellow light of twilight is the time for people and families to gather together. The black light of night is the proper time for resting. Traveling during twilight or at night is improper and increases the risk of bad things happening to a person.

The Navajo believed that following the guidelines and reminders created for them by the Holy People would lead to a prosperous and healthy life.

Awaits the Dawn is commonly known as the Milky Way. This swath of brighter sky is the view of our own flat, disc-shaped galaxy, as seen from our position within it. Even though we can't see the billions of individual stars that make up the galaxy, we can detect the increased concentration of light from them.

The Skidi Pawnee believed this band of light across the night sky was the Pathway of Departed Spirits. People who died followed this route to reach their version of heaven, where they would live in peace and be reunited with lost loved ones.

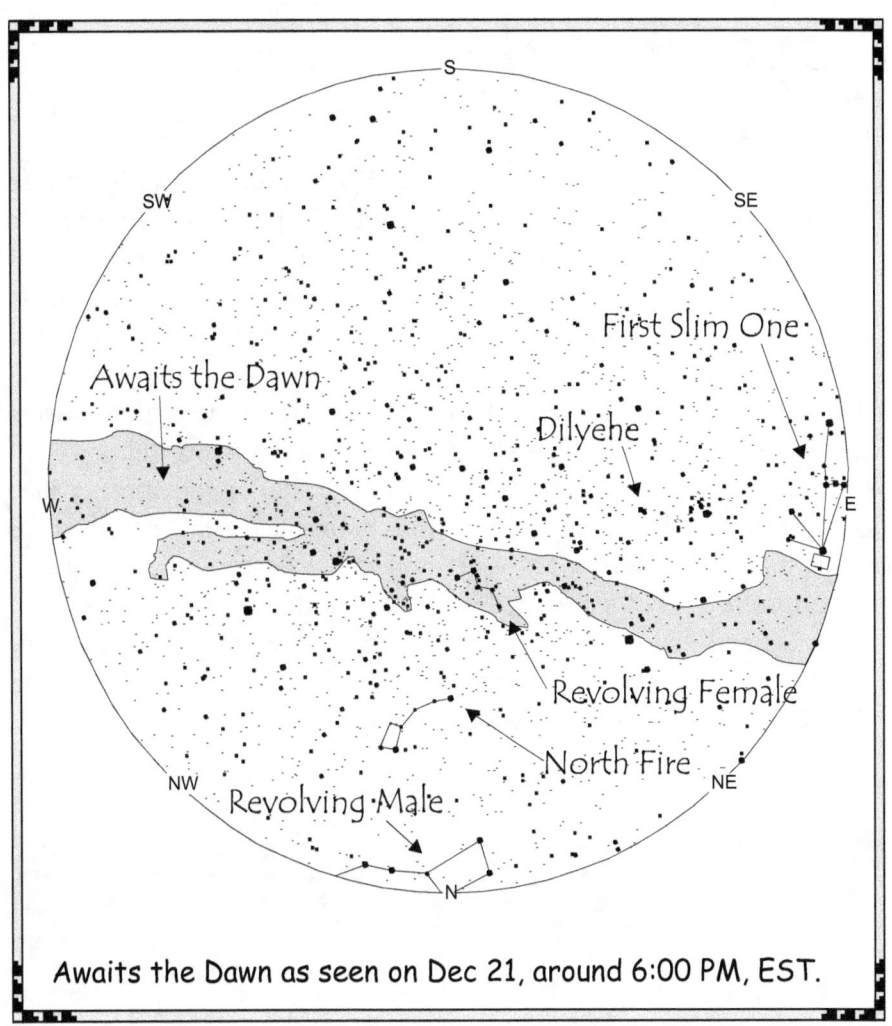

Awaits the Dawn as seen on Dec 21, around 6:00 PM, EST.

| WHERE TO LOOK 1 HOUR AFTER SUNSET | | | |
|---|---|---|---|
| WINTER | SPRING | SUMMER | FALL |
| Overhead, | Overhead | Low along E | Overhead, |
| E-W, Dec 21 | NW-SE Mar 21 | horizon | NE-SW Sep 21 |

# Coyote Star

## Mą'ii bizǫ

After Black God placed the first eight star figures in the night sky, he stopped to rest. Being a trickster, Coyote snatched the fawnskin pouch of unnamed stars from Black God.

Coyote then scattered the remaining stars at random across the sky, offsetting the orderly star figures created by Black God with a chaotic jumble of confusion.

The resulting contrast of order and chaos in the night sky represents another complementary pair in Navajo cosmology. This pairing reminds the people that their world and everything that happens in it is a balance between order and chaos.

Coyote deliberately placed the last star far to the south. This star represents the source of confusion and disorder in the world.

Coyote Star is sometimes called the "monthless star" because it is only in the sky for a short period of time, less than a month.

Coyote Star may possibly have been Canopus (-0.72, white supergiant).

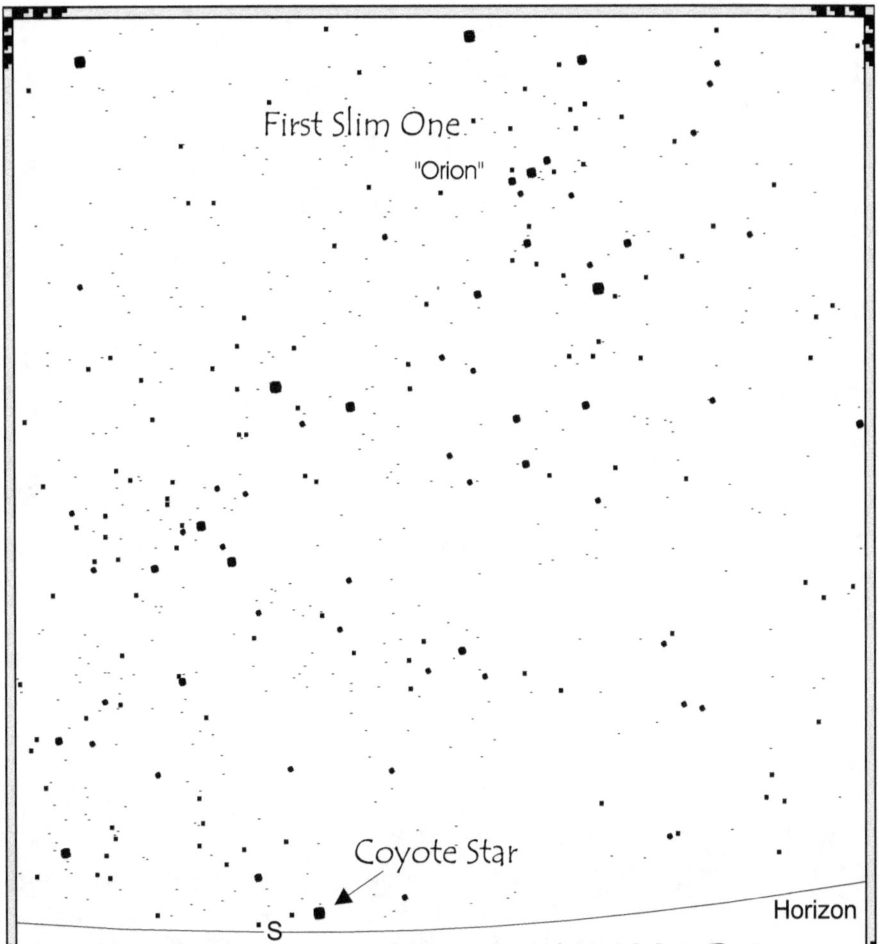

First Slim One

"Orion"

Coyote Star

S

Horizon

**Coyote Star as seen on Feb 1, around 10:00 PM, EST.**

| WHERE TO LOOK 1 HOUR AFTER SUNSET | | | |
| --- | --- | --- | --- |
| WINTER | SPRING | SUMMER | FALL |
| Below E horizon | Low in S, Mar 21 | Below NE horizon | Below NW horizon |

# Other Star Stories

Many Native American stories describe a time when animals and humans lived, worked, and talked together. As you read these stories, you'll notice references to animals and other objects, such as clouds and mountains, as living characters. While such characterization may seem cute at first, it has a more serious underlying meaning. Many Native American cultures believe that all things, whether human, animal, plant, or otherwise, have an internal presence that links us all together.

The Ojibway people, of the Great Lakes region of North America, tell a story explaining that it is Fisher we see in the group of stars now called the Big Dipper. The fisher is a small animal, similar to a large mink. It is a cunning hunter and a fierce fighter. To the Ojibway, the handle of the Big Dipper represents Fisher's long, bushy tail.

## The Birds of Summer
As retold by Don Childrey

Long ago there was a time when winter seemed like it would never end. Snow blanketed the forests and ice choked the rivers and lakes. Every day the people searched for a sign that winter was leaving, for it was time for summer's turn. But the warmth of summer did not come.

As one icy day blew into the next, the people grew more anxious and wondered what to do. Finally they called a meeting by the lodge fire. Fox, Muskrat, Beaver, and all the brothers of the tribe were there. A cunning hunter named Fisher spoke first. Fisher had traveled over many distant hills and seen many things. "Summer has not come because the birds of summer have not returned," said Fisher. .

"What could have happened to them?" the people wondered. They talked about how the birds of summer left each year as the leaves of the trees turned golden, and how they returned as the snows melted and new green leaves sprouted. "Never have the birds of summer failed to return," said the oldest of the elders.

"I think I know who might be holding back the birds of summer," Fisher said after thinking for a while. "I have a selfish cousin named Cruel-Face. He wants all good things for himself. If he has captured the birds of summer, he will try to keep them." The people responded with angry cries.

"What can we do?" they asked Fisher.

"Cruel-Face has the will of a grizzly bear," Fisher answered. "But we must go to Cruel-Face's village to bring back the birds of summer." Fisher looked slowly around the campfire at the faces of Beaver, Caribou, Fox, and the others. The people agreed.

"Prepare for a journey of ten sleeps," Fisher said. Then the people parted to gather their things.

In the morning Fisher led his people on their journey. They traveled through stands of leafless trees. They crossed wind-swept fields. They walked along trails blanketed in snow. They struggled through deep snowdrifts and crossed over rivers frozen into solid ice.

On the eighth day, winter's cold still gripped their path. All around them the world was frozen and lifeless. That night in camp, Fisher told them to look into the campfire for visions of a warm and sunny day. But some of the travelers were losing hope and mumbled about turning back.

The next morning, as the group walked along, the air began to feel warmer. Snow started to melt in the path and streams began gurgling with running water. By evening, the air had become so warm that the travelers took off their heavy winter robes. They soon reached a wide, clear lake. Sweet smelling grass grew along the shore. It felt like spring.

The travelers laid down their packs and wondered at this sudden change. Fisher pointed to the light of campfires across the lake and said, "That is the village of Cruel-Face. The breezes blow warm here because the birds of summer have never left this place. I will find Cruel-Face and release the birds of summer, but you must help me."

Then they worked out a plan. Muskrat and Beaver would start first, before dawn. They would swim across the lake and make holes in the villagers' canoes and weaken their paddles. When light of dawn began to show, Caribou would swim into the lake. Fox would then make barking noises to signal Caribou's crossing. The rest of the group would make a loud commotion that all the world could hear.

"There is one more thing," Fisher said. "If the birds of summer fly free, don't wait for me. Follow them!"

"We won't leave you behind," Caribou objected, "no matter what happens."

But Fisher insisted. "No, my friend, we must free the birds of summer. Their spirits cry out to travel their own path. You must follow them."

"Then we will see that nothing stops them!" Caribou said, and everyone agreed.

The next morning, Muskrat and Beaver swam across the lake before dawn and put holes in the villagers' canoes. When dawn came, Fox began barking and awakened Cruel-Face's people. When they crept from their lodges to see what the noise was about, they saw Caribou jumping into the lake. The villagers were eager for a hunt and raced to their canoes with their bows and arrows. Caribou swam farther out into the deep waters, encouraging the villagers to follow him. He was so far out that only his antlers could be seen from the shore.

Fisher hid in the brush and watched all of the villagers hurrying off in their canoes. Then he quietly slipped into the lodge of their chief. Once inside he found Cruel-Face seated in the center of the earthen floor. A small fire burned in front of him, making the air in the lodge thick with smoke. A pile of arrow shafts lay at his side. Surrounding him, filling the lodge from floor to ceiling, were many stacks of birchbark boxes.

The selfish chief hunched toward the fire. He was jabbing a long stick into a small heating pot sitting on glowing logs. The pot held warm sturgeon glue. Cruel-Face was using the glue to join feathers to arrow shafts.

Cruel-Face was too busy making his arrows to notice Fisher. Fisher peered through the smoky air. Then he started to creep toward the boxes.

Cruel-Face suddenly jumped up and came at Fisher like a snarling wolf. But Fisher was very quick. Snatching the glue stick away from Cruel-Face, Fisher swabbed the mouth of the chief with a thick stroke of warm glue.

This surprised Cruel-Face and his hands shot to his face, where they instantly stuck like a leech sticks to bare skin. He tried to cry out, but no sound came out. His mouth was glued tightly shut. Cruel-Face thrashed around the lodge in a rage. Feathers flew and boxes crashed and sticks scattered.

Fisher wasted no time. He dashed swiftly among the boxes, opening one here, poking a hole in another there. Just as he had suspected, Cruel-Face had captured the birds of summer and was keeping them in these boxes. From the first box flew the thrushes and the warblers. From the

second came finches and sparrows. Then came jays and wrens, swallows, woodpeckers, and bluebirds.

While his cousin hopped angrily about the lodge, Fisher continued opening every box he could find. A cloud of birds rose like a rainbow from the lodge of the selfish chief. Just like smoke, it drifted out and over the wide lake. The birds flew over the villagers paddling in their now leaking canoes.

"Look!" cried one of Cruel-Face's warriors.

"Listen!" called others. "The birds of summer are getting away!" They urgently began to turn their canoes around toward the village. Their paddles, damaged by Muskrat and Beaver, began to break under the panicked strokes of the villagers. The canoes soon took on more water and sank. The people of Cruel-Face were left swimming for shore.

The birds of summer continued to fly across the lake and over the forests. Fox, Caribou, Muskrat, Beaver, and all of Fisher's friends gathered on the grassy shore and quickly began to follow the birds of summer. Behind them a winter wind was starting to swirl around the lodges of Cruel-Face's village.

Fisher was still dodging the angry chief as he chased him around the lodge, but one small box remained to be opened. Using an arrow shaft, Fisher was able to poke a small hole in the box and out darted a flurry of hummingbirds. While they flashed and circled around the head of Cruel-Face, Fisher dashed out of the lodge.

Fisher ran like a rabbit running from a fox to rejoin his friends. He ran through Cruel-Face's village, past the lodges, and right into the angry crowd of cold, wet villagers emerging from their unintended swim in the lake.

Fisher crouched like a cornered animal, looking for an escape. Then he sprang towards a tall tree. Climbing swiftly, Fisher could hear the shouts and hard breathing of the angry villagers climbing up behind him. The tree shook and branches snapped. Soon he reached the top of the tree. Above him stretched only the sky.

"Brave Fisher!" whispered the stars. "Brave Fisher, you are our brother." The voices of the stars called like the song of a hundred birds. Fisher reached out with his arms. Then, slipping from the grasp of the angry people, Fisher rose over treetops and hills, joining the friendly stars in the winter sky.

Beaver, Caribou, Fox, and the rest of the travelers followed the birds of summer towards to their village. As they traveled, they saw flowers bloom and leaf buds unfold. But fear filled their hearts when the people thought about what might have happened to Fisher. Before they made it back to their own village, they decided to turn back on the path again to search for their friend.

As they gathered around the campfire one night, a small boy pointed to the sky. "I see Fisher there among the stars!" he cried. The people looked up excitedly.

"Brave Fisher has escaped to the sky country!" they murmured to one another. Then they hurried back to their village to tell the story of Fisher and the birds of summer to all the people of their village.

"We will keep the birds of summer for ourselves from this day on," some of the villagers said when they heard the story. "We can make sure the warm winds of summer always blow across our land."

But others looked up at the sky and asked, "What would Fisher say about this?"

"Fisher would say the birds of summer should always be free to travel their own path," said Fox, and many agreed.

That is why things are as they are. For half of the year the people walk happily in warm breezes. They enjoy the fragrance of a thousand flowers and listen to the songs of the birds of summer. But when the traveling moon appears in the heavens, the birds of summer take wing. Cold winds blow and snow covers the path. The sun sinks lower in the sky. That is when the people search the winter sky for Fisher and make the long nights pass more quickly by telling stories about their brave friend.

We call the pattern of seven bright stars that appears to swing continuously around the North Star the Big Dipper, because its shape reminds us of a dipper or large ladle. In ancient Greece it was known as Ursa Major, the Great Bear. Many Native American people also associated this group of stars with the figure of a bear. The Onondaga, the Cherokee, the Blackfoot, the Zuni, and some Eskimo groups, as well as the Micmac of eastern Canada, thought of the Big Dipper as a bear pursued by hunters. This celestial bear is visible all year long but looks like it is standing upright in some months and upside down in others. The Micmac people told a story to explain the changing look of the night sky. It is a story that never ends.

### The Never-Ending Bear Hunt
As retold by Don Childrey

Far away in the sky country, the warm breezes of spring were beginning to melt the snow and ice of winter. New leaves and green plants were popping up everywhere. A bear woke up from her long winter sleep. After such a long nap, she was hungry. She left her rocky den on the hill and lumbered off to search for berries.

In the bushes nearby, a sharp-eyed chickadee was flitting around. When he saw the bear, he  quickly flew off to gather  his friends around him.

"It has been a long winter and I am hungry," said Chickadee. "I just saw a bear nearby and I have decided to go bear hunting. Who will come with me?"

"We are all hungry too," said Moose-bird. "I will go with you." Five more of his friends agreed to join Chickadee.

Robin was chosen to lead the hunting party and he took his place at the head of the line. Following him was Chickadee, who was the smallest hunter but also the wisest. He carried along a pot for boiling his dinner. Moose-bird was impatient and fell in line behind Chickadee. He intended to help himself to Chickadee's pot. Behind Moose-bird were Pigeon, Blue

Jay, Horned Owl and Saw-whet. As soon as the hunting party was ready, they set out to find the bear.

Food had been scarce through the winter, so the hunters hurried as they followed the bear's tracks. The bear appeared to be large and awkward and they thought they would soon catch up with her. But no matter how fast they traveled, they could not catch up with her. The bear continued to stay ahead of the hunters.

All through the summer the bear traveled across the sky country in search of berries. The hungry hunters followed her the whole way. As summer passed into autumn, the hunters became tired and some of them began to lose interest in chasing the bear.

"I'm too far behind to catch up," said Saw-whet at the back end of the line. He was a clumsy owl and slower than the other hunters. Saw-whet was the first one to drop out of the chase.

"Where has that lazy Saw-whet gone?" asked Horned Owl. He soon left to go search for his cousin.

Blue Jay tried to keep up, but eventually he and Pigeon fell back too far and lost their way. . Only Robin, Chickadee with his pot, and Moose-bird were left to follow the trail of the bear.

At last, well into autumn, the bear had eaten so many berries that she began to slow down. The hunters were then able to catch up with her. They cornered her against a huge rock. The bear reared up on her hind legs and tried to defend herself. She was growling and clawing as she stood, but Robin shot her with his arrow. The bear fell over on her back. Her blood spilled out and colored the leaves of autumn with many shades of red. Her blood also stained Robin's breast a brilliant red.

For the rest of the winter, the bear lay dead on her back in the sky country. But that was not the end of the bear's story. Her life spirit entered another bear who had found the rocky den on the hill and was sleeping away the winter. When the warm breezes of spring arrive again and awaken the sleeping earth, the den will reappear. A new star bear will come out of the den to search for berries in the sky, and Chickadee will begin the hunt again, as he does every year.

We know the bright swath of stars stretching across the night sky as the Milky Way. Among Native American tribes, there were nearly as many different stories about the Milky Way as there were tribes. Some believed it was the path that birds followed in their migrations to the south. The Pawnee, who lived in the Nebraska area, believed it was the pathway taken by departed spirits as the traveled to the Southern Star, where they would live for eternity. The Snohomish, who lived in the Puget Sound region, were whalers and fishermen who saw this feature in the sky as a river and the scene of a conflict between the people of earth and the people of the sky.

### Adventure Along The Great White River In The Sky
As retold by Don Childrey

Long ago, when the earth was cold and dark, there was a man who made canoes better than anyone else. His canoes were so tight and strong that they rode the water like sleek silver fish. The canoemaker loved his work so much that he would rise every day before dawn, before the stars had left the sky, to begin work. He would work all day and late into the night. He hammered and pounded, scraped and rubbed, until his fine canoes were capable of handling the angriest waters.

Up in the heavens, the great chief of the sky people could hear the canoemaker's hammering. Day after day, the pounding and scraping went on. The noise began to irritate the great chief. When he reached the point where he could not stand the noise any more, the great chief sent down four of his men to capture the canoemaker.

The people of the canoemaker's village were proud of the canoemaker and his skills. No one else could make canoes as fine as his. When the villagers discovered that he was gone, they began searching everywhere for him. They looked under the unfinished canoes, in the forests, along the streams, and across the frozen hills. . Finally, they found an old man who spent his sleepless hours looking out his window. He told the villagers

that he had seen four sky people capture the canoemaker and take him off to the sky country.

The villagers were very concerned and upset by this news. "What will we do without our canoemaker?" they wailed.

While the villagers were crying about their loss and wondering what to do, the old man spoke up. "When I was a boy, storytellers used to say the sky could be reached by making a chain of arrows."

The villagers didn't have any other ideas, so they took up their bows and started shooting a flurry of arrows into the sky. The volley of arrows up towards the heavens like a swarm of mosquitoes, but none of them went high enough to stick in the sky.

Robin, who had been flying about from tree to tree and watching the villagers, finally came forward to offer his help. He took an arrow and flew up to the sky. He was able to stick the first arrow into the heavens. Then the canoemaker's people added theirs onto this first arrow. Finally a chain of arrows extended from the sky all the way down to the village.

Then the villagers began to climb the arrows into the heavens. They were careful to be very quiet so the great chief of the sky people would not hear them.

When the last of the villagers had climbed into the sky, Robin began to fly around the sky country, looking for the canoemaker. Soon he found the village of the sky people. It was located along the Great White River in the Sky. Then he found the canoemaker tied to the roof of one of the houses. Robin quietly told the canoemaker that help was coming, and then he raced back to alert the villagers.

It was cold in the sky country, and but Robin continued to snoop around some more. As he was looking around the sky country, Robin noticed a warm feeling and went to investigate. Near the Great White River, he found a group of sky people gathered around a fire. Since the lower world had not been given fire yet, Robin had no idea what this strange glowing thing was. But he knew the warmth felt good! "Whatever this thing is, we should have some of it in our village," Robin said to himself.

Moving quietly, Robin tried to blend in with the sky people and move closer to the fire., But each time he came near, the sky people around the fire would crowd together and push him back into the cold. Robin was angry and frustrated. He left to go find his friend Beaver. He told Beaver about the fire and it's warmth. Together they came up with a plan.

Beaver swam out into the Great White River and then floated down to the place where the sky people were gathered around the fire. Some of the children saw Beaver and began to chase and tease him. This was part of the plan. Beaver played the children's game for a while, but as he ran around he worked his way closer and closer to the fire.

Finally Beaver was able to get very close to the fire. Just at that time, the people from canoemaker's village burst into camp to rescue their friend. The sky people jumped up with a roar and began chasing after them.

In all the confusion, Beaver saw his chance. He ran quickly to the fire, snatched a flaming stick, and then ran off to the hole in the sky as fast as he could.

Robin led canoemaker's people to the rooftop where they freed their friend. Then all of the villagers  ran along the shore of the Great White River until they reached the hole in the sky. By this time, the chief of the sky people had heard all of the noise. He came thundering after them. But canoemaker and his people reached the chain of arrows and climbed out of the sky before he could reach them..

Safely back in the canoemaker's village, the people warmed themselves around the fire that Beaver brought back from the sky country. The canoemaker was happy to be home. He thanked the villagers and promised, "I will never make the great chief of the sky people angry again." That is why canoemaker and his people never work on their canoes early in the morning or late at night.

# Helpful Charts and Data

## Sky Viewing Guides

The Sky Viewing Guide charts tell you whether the phase and brightness of the moon should affect visibility on the date you will be out viewing. You will also find indicators for dates of the major meteor showers on these charts.

To supplement these guides, you will need to check other sources* for the moon's rise and set times for your specific viewing location. If the moon is full and above the horizon during the time you will be viewing, it's light will hide most stars and meteor trails. If the moon is below the horizon, you will have the best viewing opportunity.

Visibility is also dependent upon the sky being free of clouds or haze, and your viewing location being away from  sources of light pollution.

|  |  |
|---|---|
| ⊛ | Full Moon |
| ◐ | Waning Quarter Moon |
| ● | New Moon |
| ◑ | Waxing Quarter Moon |
| **Q** | Quadrantid Meteor Shower |
| **L** | Lyrid Meteor Shower |
| **E** | Eta Aquarid Meteor Shower |
| **D** | Delta Aquarid Meteor Shower |
| **P** | Perseid Meteor Shower |
| **O** | Orionid Meteor Shower |
| **T** | Taurid Meteor Shower |
| **Le** | Leonid Meteor Shower |
| **G** | Geminid Meteor Shower |

(For more information on meteor showers, see the Meteor Shower section.)

*One source for moon rise/set times is the US Naval Observatory's website: http://aa.usno.navy.mil/data/docs/RS_OneDay.html

## How to use the Sky Viewing Guide Charts
Step 1: Find the chart for the correct year.
Step 2: Find the month on the top side of the chart.
Step 3: Move down the column to the row for the specific date, noting the symbols you encounter along the way.

## Example for May 5, 2004:

**1:** Find the chart for 2004.

**2:** Find the month of May on the top side of the chart.

**3:** Moving down the column to the row for the fifth day, note the following events:
       - Full Moon on the 5th
       - Meteor shower on the fifth

With clear skies on the evening of May 5, 2004, you should be able to see a full moon. Check other sources for the moon rise time at your location. You may also be able to see the Eta Aquarid meteor shower. If the moon has risen, it's light may hide many stars and meteor trails.

## For more information

The US Naval Observatory website offers several tools for predicting moon rise and set times, as well as other astronomical events:

http://aa.usno.navy.mil/data/

## Example:

# 2004 Sky Viewing Guide

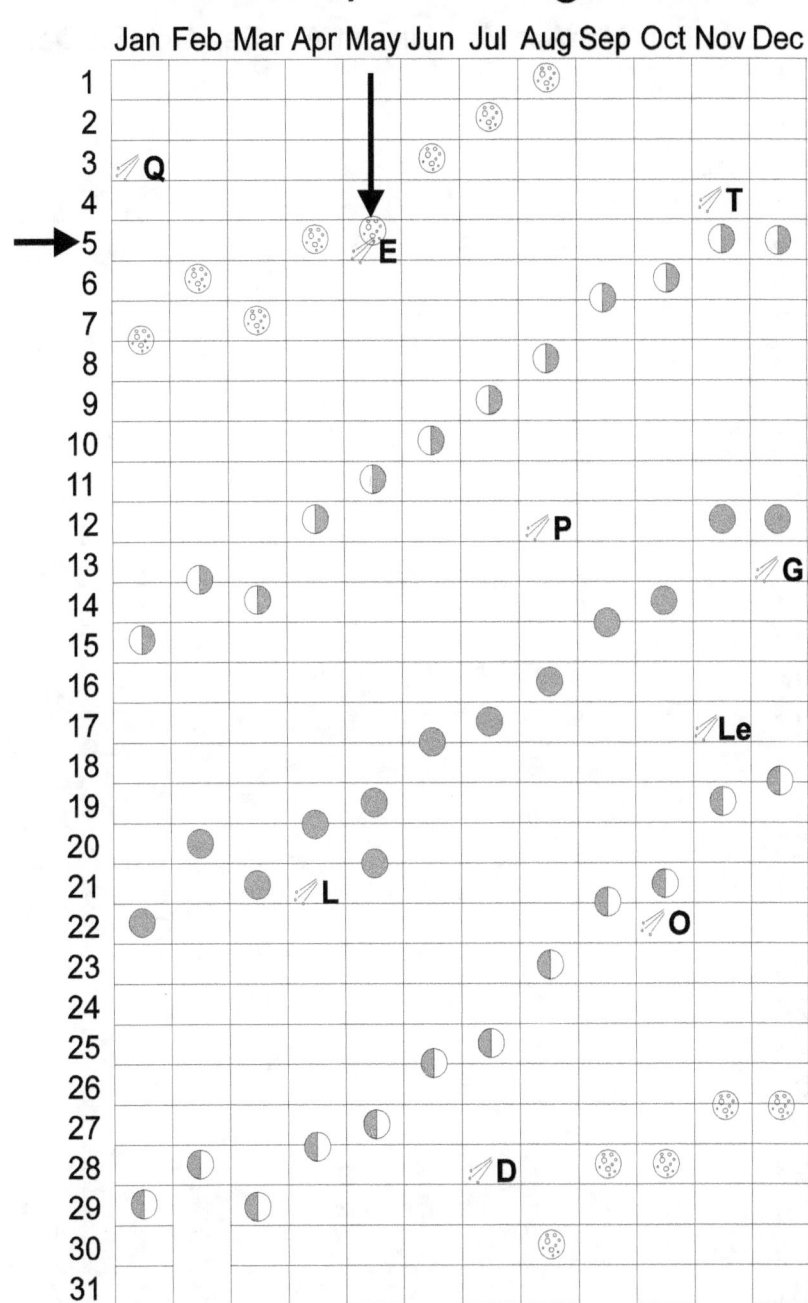

# 2009 Sky Viewing Guide

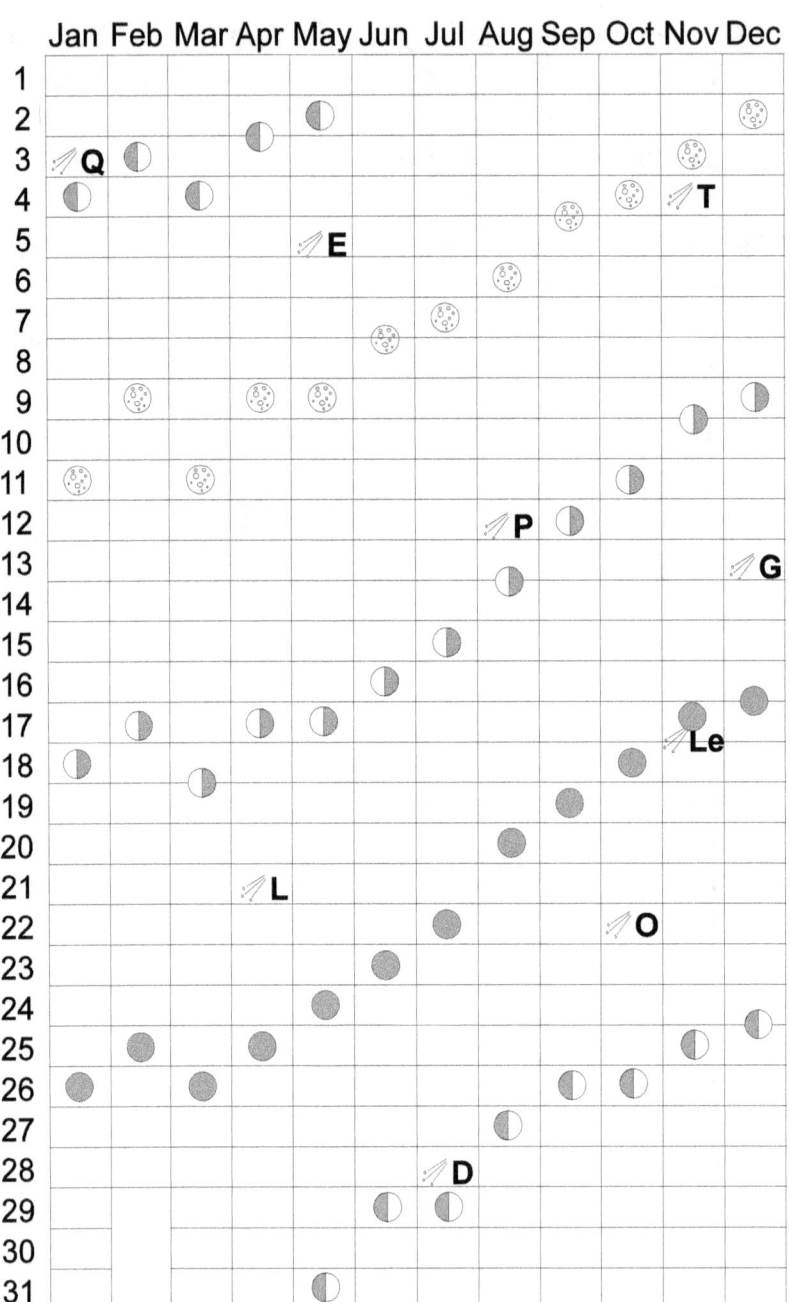

# 2010 Sky Viewing Guide

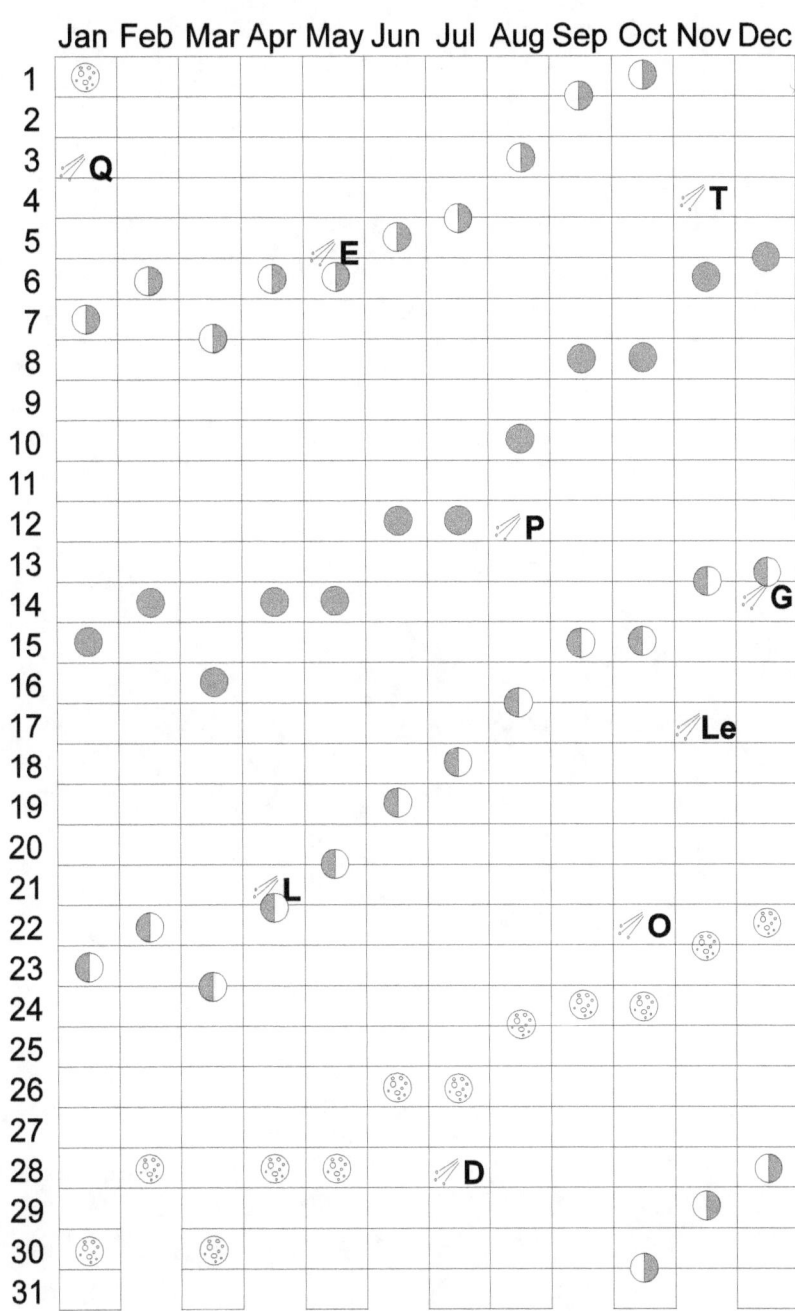

## Planet Visibility Charts

The Planet Visibility Charts will help you identify which of the four planets: Venus, Mars, Jupiter and Saturn, should be visible on a given date and time of day. These planets are easily spotted with the naked eye. Mercury is sometimes visible to the naked eye, but it is much more difficult to catch the right conditions for viewing it, so I have not included it in these charts.

By looking up the date when you will viewing, these charts should quickly tell you which planets should be overhead at the time, and where they will be on their path across the sky. The rise and set times are for a location in central North Carolina, but should be relatively close for most other locations.

Each chart covers one year. The vertical scale represents the days of the year, while the horizontal scale represents the time of day. By locating the appropriate date and time on the chart, and then noting the planet rise and set lines, you should be able to tell which planets will be above the horizon and visible at that time.

The period of time between sunset and true darkness is indicated on the charts as a shaded band on the left side of the chart. Likewise, the period of time between true darkness and sunrise is also shown as a similar shaded band on the right side of the chart.

The periods of evening and morning twilight provide an excellent time to spot these planets. The stars are usually hidden by the light of the sky at this time, but the planets are not. Because they are significantly closer, the planets are much brighter and will be visible when stars are not.

Visibility of the planets is also dependent upon the sky being free of clouds or haze, and your viewing location being away from manmade sources of light pollution.

**How to use the Planet Visibilty Charts**
Step 1: Find the chart for the correct year.
Step 2: Find the date on the left side of the chart.
Step 3: Move to the right across the chart, noting the significance of each line you cross.

**Example:**
**1:** Find the chart for 2004.

**2:** Find May 1 on the left side of the chart.

**3:** Moving to the right across the chart, note the following events:
- 6:45 PM, sunset
- 8:40 PM, true darkness
- 11:30 PM, Venus and Mars set
- 12:30 AM, Saturn sets
- 3:30 AM, Jupiter sets

If you are out viewing on the evening of May 1, 2004, you should be able to see Venus, Mars, Saturn and Jupiter in the western sky. Because they will have risen prior to sunset, they should be the first "stars" you see that evening. Because they all set before sunrise, none of them will be visible at dawn the next morning.

**For more information**

The US Naval Observatory website offers several tools for predicting planet rise and set times, as well as other astronomical events:

http://aa.usno.navy.mil/data/docs/mrst.php

## Example:

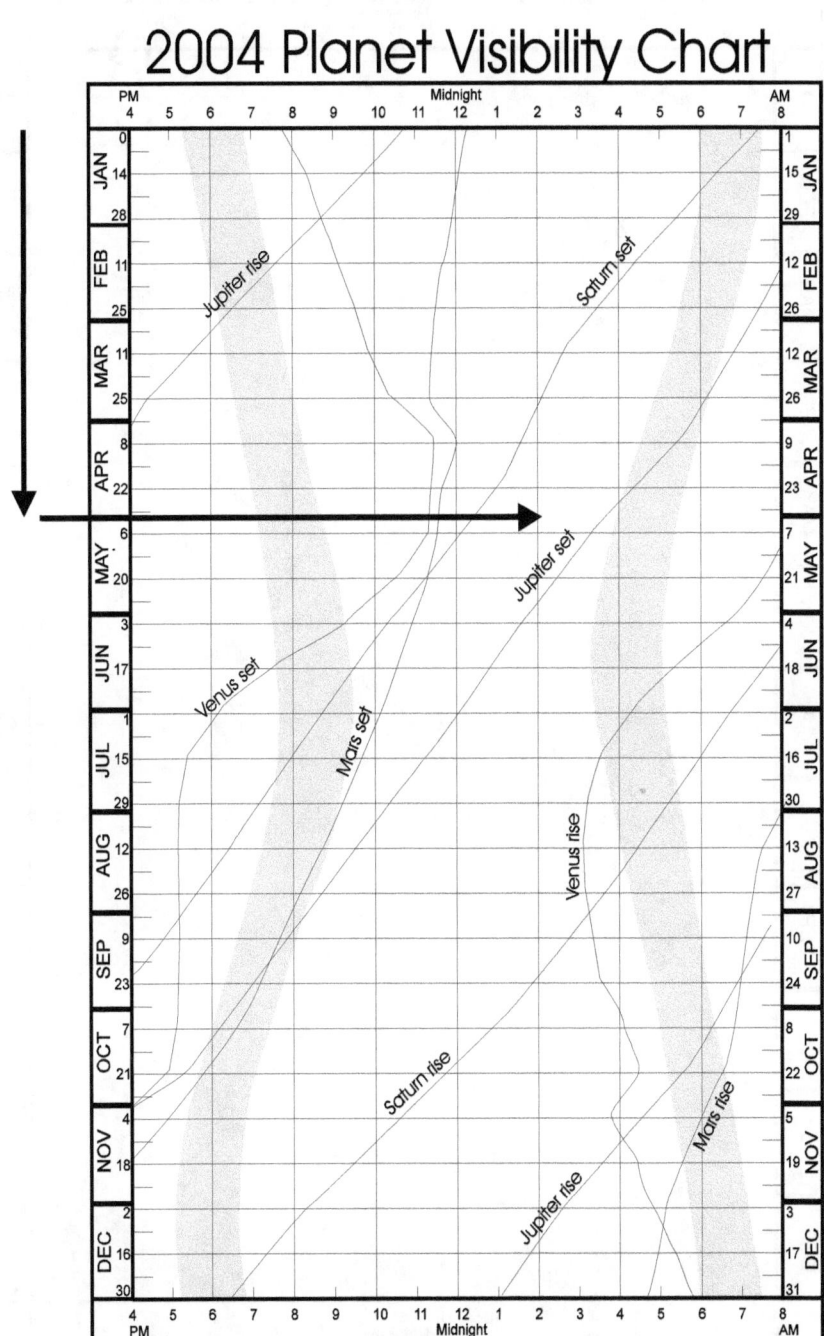

# 2004 Planet Visibility Chart

# 2009 Planet Visibility Chart

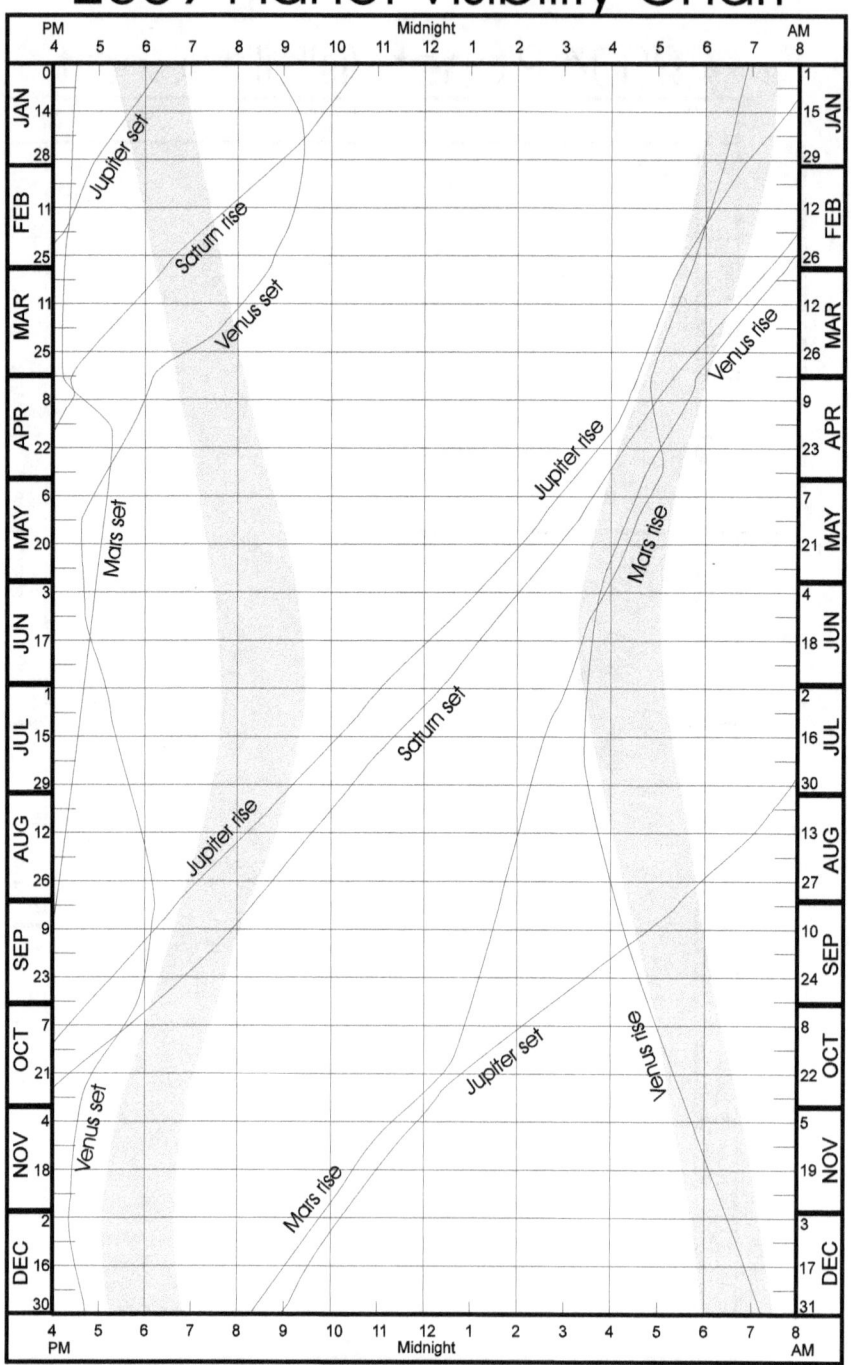

# 2010 Planet Visibility Chart

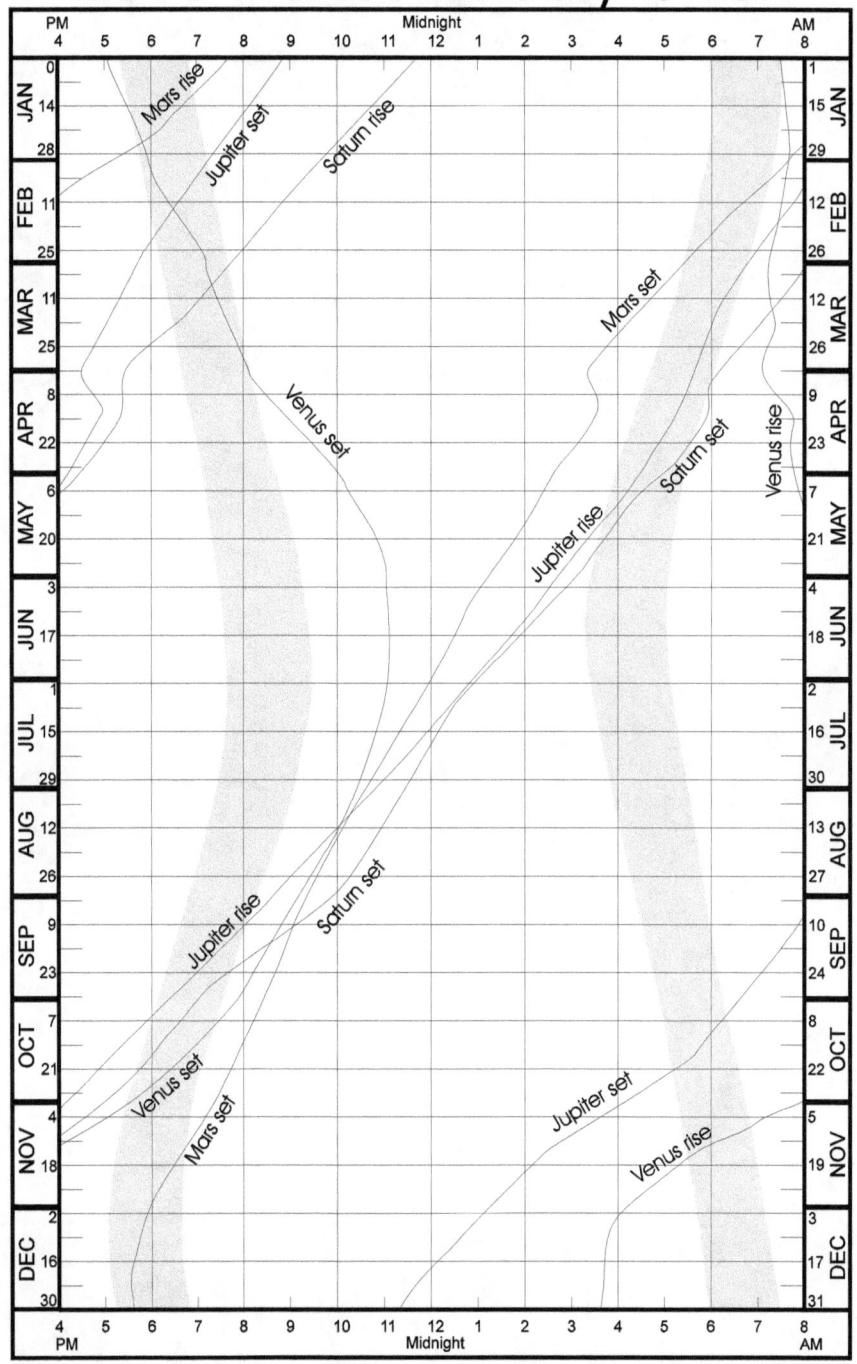

## Meteor Showers

Meteor showers are the result of the earth passing through a stream of debris left in space by passing comets, asteroids, or other objects. Because the earth follows a predictable path through space each year, astronomers can tell us when these meteor showers will occur.

Meteor showers may sometimes last for several days, depending on the width of the source debris field. Most showers will have a peak, or maximum, during which the most meteors will be seen. An average rate of meteors per hour at the maximum is listed in the chart below for each shower. The actual rate seen often varies.

As our planet moves through space in it's orbit around the sun, the side where "dawn" is occurring is the "leading" side. This is also the side running into the debris in space that becomes our meteors. If you want to view meteor showers, the best time is between between midnight and dawn. The early evening hours are on the "trailing" side of the planet, which generally means less meteor visibility.

Meteor showers are named for the constellation in which their apparent point of origin, or radiant, appears to be located.

Visibility of meteor showers is also dependent upon the sky being free of clouds or haze, and your viewing location being away from sources of light pollution.

| Meteor Shower | Date of Maximum | Radiant Location and Source | Avg. Rate/Hr |
|---|---|---|---|
| Quadrantids | 3-Jan | NE | 10-50 |
| Lyrids | 22-Apr | NE    Comet 1861 I Thatcher | 5-15 |
| Eta Aquarids | 4-May | E    Comet Halley | 10-20 |
| Delta Aquarids | 28-Jul | SE | 10-20 |
| Perseids | 11-Aug | NE    Comet 1862 III Swift-Tuttle | 30-70 |
| Orionids | 21-Oct | E    Comet Halley | 10-30 |
| Taurids | 3-Nov | SW    Comet Encke | 10-20 |
| Leonids | 17-Nov | E    Comet Temple-Tuttle | 10-20 |
| Geminids | 13-Dec | E    Asteroid 3200 Phaethon | 30-80 |
| Ursids | 22-Dec | N    Comet Tuttle | 10-20 |

Chart sources: Peterson Field Guide Stars and Planets, 1992; Summer Stargazing, 1996

## Full Moon Names

You may have heard of the Harvest Moon as a name for the full moon in September, but did you know that there are names for each month's full moon? Many of these names are derived from Native American tribes, especially those of the northern and eastern US. In fact, these tribes referred to the whole month by these names, and used them to help keep track of the seasons.

**Wolf Moon** - January
The cold and snow of midwinter caused hungry wolf packs to prowl around Native American villages. Some tribes called this the *Snow Moon*.

**Snow Moon** - February
The heavy snows of February lend their name to this moon. Because hunting was so difficult due to the snow, some tribes called this the *Hunger Moon*.

**Worm Moon** - March
This is the month that the ground begins to soften and earthworms become active again, inviting the return of the robins. Some tribes considered the cawing of the crows to signal the end of winter and called this the *Crow Moon*. Some tribes called it *Crust Moon*, because the warmer days and freezing nights created a hard crust on the snow. It is also known as the *Sap Moon*, indicating the time of year to begin tapping maple trees.

**Pink Moon** - April

Grass pink phlox, or wild ground phlox, is one of the first flowers to bloominthe spring, thereby lending it's name to this moon. some tribes called it the *Sprouting Grass Moon*. Coastal tribes called it the *Fish Moon*, because the shad would come upstream to spawn at this time.

**Flower Moon** - May

The flowers of May lend their name to this moon, also known as the *Corn Planting Moon*.

**Strawberry Moon** - June

All of the tribes of the Algonquin nation knew this moon as the *Strawberry Moon*. It signaled the time to harvest the first strawberries.

**Buck Moon** - July

This is the month that new, velvet-covered antlers begin to show on male deer. Some tribes knew this as *Thunder Moon*, becuase thunderstorms are so common during this month. Tribes along the Great Lakes knew this as the *Sturgeon Moon*, because these large fish are most readily caught at this time. Other tribes knew this as the *Hay Moon, Red Moon, Green Corn Moon* or the *Grain Moon*.

**Fruit Moon** - August

Also known as the *Barley Moon*. This name is usually only used when the Harvest Moon falls in late September or early October.

**Harvest Moon** - September

The Harvest Moon is defined as the Full Moon closest to the Autumnal Equinox. Every third year, this full moon is in October. Normally, the moon rises about 50 minutes later each day, but for a few nights near the Harvest Moon, it rises within 10 to 20 minutes of the same time, giving farmers the chance to work into the night gathering their harvests. The primary Native American crops - corn, pumpkins, squash, beans, and wild rice are ready for harvest at this time.

**Hunter's Moon** - October

With the animals fattened up by the fall harvest of nuts and fruits, this is the best time of the year for hunting. This was also known as the *Traveling Moon*, or *Dying Grass Moon*.

**Beaver Moon** - November

This is the time of the year when beavers become more active in preparation for winter. This is also the time to set beaver traps before the swamps freeze over for the winter. This moon is sometimes called the *Frosty Moon*.

**Cold Moon** - December

Also known as the *Long Nights Moon*, because the longest nights of the year are in this month, at the time of the Winter Solstice.

**Blue Moon** -

This is the name that people of today give to the second full moon that occurs in a single calendar month. This usually only happens once every three years.

Sources:

http://www.almanac.com/details/moonnames.php

http://www.space.com/spacewatch/full_moon_names_040102-1.html, by Joe Rao, SPACE.com's Night Sky Columnist, January 2, 2004

**Seasonal Guide to Star Figures**

The following guides will help you know what to look for each season.

**Spring** - Defined today as the period from Vernal Equinox to Summer Solstice (approximately March 21 to June 21). The Vernal Equinox occurs when the sun's apparent path around the earth crosses the equator, as it is progressing northward. The hours of daylight equal the hours of nighttime.

The Navajo divided their year into just two seasons - Summer and Winter. Summer lasted from March into October.

**Summer** - Defined today as the period from Summer Solstice to Autmnal Equinox (approx. June 21 to Septemer 21). The Summer Solstice occurs when the sun's apparent path around the earth has reached it's northernmost limit. This is the longest day (sunrise to sunset) in the northern hemisphere.

**Fall** - Defined today as the period from Autumnal Equinox to Winter Solstice (approx. September 21 to December 21). The Autumnal Equinox occurs when the sun's apparent path around the earth crosses the equator, as it is progressing southward. The hours of daylight equal the hours of nighttime.

For the Navajo, the Winter season began in October and lasted until March.

**Winter** - Defined today as the period from Winter Solstice to Spring Equimox (approx. December 21 to March 21). The Winter Solstice occurs when the sun's apparent path around the earth has reached it's southernmost limit. This is the shortest day (sunrise to sunset) in the northern hemisphere.

# Spring March 21 - June 21

At one hour after sunset:

**North Fire** - Steady in the northern sky.

**Revolving Male** - High in the northeastern sky, with his head towards the east and his feet towards the southwest.

**Revolving Female** - High in the northwestern sky, with her head towards the south and her feet towards the northwest.

**Dilyehe** - Midway up in the western sky.

**First Slim One** - Midway up in the southwestern sky, with his feet toward the western horizon and his head up.

**First Big One** - Not visible; waiting to rise in the east.

**Man With Feet Apart** - Rising in the southeastern sky, laying along the horizon with his feet toward the east and his head toward the south.

**Rabbit Tracks** - Not visible; waiting to rise.

**Awaits the Dawn** - Overhead, west of the meridian, running NW to SE.

**Lyrid meteor shower** occurs around April 22.

**Eta Aquarid meteor shower** occurs around May 4.

# Summer June 21 - September 21

At one hour after sunset:

**North Fire** - Steady in the northern sky.

**Revolving Male** - High in the northwestern sky, with his head up and his feet down towards the northern horizon.

**Revolving Female** - Low along the northeastern horizon, with her head towards the west and her feet towards the east.

**Dilyehe** - Not visible; below the northwestern horizon.

**First Slim One** - Not visible; below the northwestern horizon.

**First Big One** - Rising up in the southeastern sky, with his head up and feet down toward the southeastern horizon.

**Man With Feet Apart** - Setting in the southwwestern sky, laying along the horizon with his feet toward the south and his head toward the west.

**Rabbit Tracks** - Rising in the southeastern sky.

**Awaits the Dawn** - Low along the eastern horizon.

**Delta Aquarid meteor shower** occurs around July 28.

**Perseid meteor shower** occurs around August 11.

# Fall September 21 - December 21

At one hour after sunset:

**North Fire** - Steady in the northern sky.

**Revolving Male** - Low in the northwestern sky, with his head towards the west and his feet down towards the north.

**Revolving Female** - High in the northeastern sky, with her head down towards the northern horizon and her feet up.

**Dilyehe** - Rising in the northeast.

**First Slim One** - Not visible; below the northeastern horizon.

**First Big One** - Low along the southwestern horizon, with his head towards the west and his feet towards the south.

**Man With Feet Apart** - Not visible; below the western horizon.

**Rabbit Tracks** - Very low along the southwestern horizon.

**Awaits the Dawn** - Overhead, running NE to SW.

**Orionid meteor shower** occurs around October 21.

**Taurid meteor shower** occurs around November 3.

**Leonid meteor shower** occurs around November 17.

**Geminid meteor shower** occurs around December 13.

# Winter December 21 - March 21

At one hour after sunset:

**North Fire** - Steady in the northern sky.

**Revolving Male** - Low in the northern sky, with his head to the west and his feet towards the east.

**Revolving Female** - High in the northern sky, with her head towards the east and her feet towards the west.

**Dilyehe** - High in the southeastern sky.

**First Slim One** - Rising in the east, with his head towards the north and his feet towards the south.

**First Big One** - Not visible; below the western horizon.

**Man With Feet Apart** - Not visible; below the northern horizon.

**Rabbit Tracks** - Not visible; below the southwestern horizon.

**Awaits the Dawn** - Overhead, running E to W.

**Quadrantid meteor shower** occurs around January 3.

**Ursid meteor shower** occurs around December 22.

## Star Types

**Star Colors (Spectral Type)**

| Color | Temperature | Example | Navajo star figure | Constellation |
|-------|-------------|---------|--------------------|----------------|
| Blue | 10,000 - 45,000 °K | Rigel | First Slim One | Orion |
| Blue-White | 7,500 - 10,000 °K | Ruchbah | Revolving Female | Cassiopeia |
| White | 6,000 - 7,500 °K | Polaris | North Fire | Ursa Minor |
| Yellow-White | 5,000 - 6,000 °K | Muscida | Revolving Male | Ursa Major |
| Orange | 3,500 - 5,000 °K | Schedar | Revolving Female | Cassiopeia |
| Orange-Red | 2,500 - 3,500 °K | Antares | First Big One | Scorpius |

Note: Our Sun is a Yellow-White star.

**Star Sizes (Luminosity Classes)**

| Size | | Example | Navajo star figure | Constellation |
|---|---|---|---|---|
| Supergiants | Up to 1000x Sun | Betelgeuse | First Slim One | Orion |
| | | Rigel | First Slim One | Orion |
| Bright Giants | | Polaris | North Fire | Ursa Minor |
| | | Antares | First Big One | Scorpius |
| Giants | 50x Sun | Bellatrix | First Slim One | Orion |
| | | Tania Australis | Revolving Male | Ursa Major |
| Sub Giants | | Tania Borealis | Revolving Male | Ursa Major |
| | | Shaula, Lesath | First Big One | Scorpius |
| Main Sequence Stars | Similar to Sun | Ruchbah | Revolving Female | Cassiopeia |
| | | Sun | - | - |
| White Dwarfs | Similar to Earth, 1/100x Sun | Proxima Centauri | - | Centaurus |

Note: The distance between a star and earth can make larger stars appear to be smaller.

## 10 Brightest Stars ( Apparent brightness)

| Name | Magnitude | Navajo star figure | Constellation |
|---|---|---|---|
| Sun | | | |
| Sirius | -1.46 | | Canis Major |
| Canopus | -0.72 | | Carina |
| Arcturus | -0.04 | | Bootes |
| Alpha Centauri | -0.01 | | Centaurus |
| Vega | 0.03 | | Lyra |
| Capella | 0.08 | | Auriga |
| Rigel | 0.12 | First Slim One | Orion |
| Procyon | 0.37 | | Canis Minor |
| Betelgeuse | 0.41 | First Slim One | Orion |
| Achernar | 0.46 | | Eridanus |

Note: The lower magnitude numbers represent greater brightness.

# Bibliography

Chamberlain, Von Del. 1982. *When Stars Came Down To Earth, Cosmology of the Skidi Pawnee Indians of North America.* Ballena Press Anthropological Papers No. 26. ISBN 0-87919-098-1.

Chartrand, Mark R. 1998. *National Audubon Society Field Guide to the Night Sky.* Alfred A. Knopf, New York. ISBN 0-679-40852-5.

Dickinson, Terence. 1996. *Summer Stargazing, A practical Guide For Recreational Astronomers.* Firefly Books, Buffalo, New York. ISBN 1-55209-047-7.

Gibson, Stephen. "The Pleiades". HTTP://www.ras.ucalgary.ca/~gibson/pleiades/

Griffin-Pierce, Trudy. 1992. *Earth Is My Mother, Sky Is My Father; Space, Time, and Astronomy in Navajo Sandpainting.* University of New Mexico Press, Albuquerque. ISBN 0-8263-1634-4.

Haile, Berard. 1947. *Starlore Among The Navajo.* Museum of Navajo Ceremonial Art, Santa Fe, New Mexico.

Mayo, Gretchen Will. 1987. *Star Tales; North American Indian Stories about the Stars.* Walker and Company, New York.

Miller, Dorcas. 1997. *Stars of the First People; Native American Star Myths and Constellations.* Pruett Publishing Company, Boulder, Colorado. ISBN 0-87108-858-4.

Monroe, Jean Guard and Ray A. Williamson. 1987. *They Dance in the Sky; Native American Star Myths.* Houghton Mifflin Company, Boston.

Pasachoff, Jay M. and Donald H. Menzel. 1992. *A Field Guide to the Stars and Planets.* The Peterson Field Guide Series. Houghton Miflfin Company, New York. ISBN 0-395-53759-2.

Rao, Joe. SPACE.com HTTP://www.space.com/spacewatch/full_moon_names_040102-1.html, by Joe Rao, SPACE.com's Night Sky Columnist, January 2, 2004

Rey, H. A. 1976. *The Stars; A New Way To See Them.* Houghton Mifflin Company, Boston. ISBN 0-395-24830-2.

Ridpath, Ian. 1995. *Stars and Planets.* Collins Pocket Guide. Harper Collins Publishers, London. ISBN0-00-219979-3.

Staal, Julius D. W.1988. *The New Patterns in the Sky; Myths and Legends of the Stars.* The McDonald and Woodward Publishing Company, Blacksburg, Virginia. ISBN 0-939923-04-1.

Williamson, Ray A. and Claire R. Farrer. 1992. *Earth and Sky; Visions of the Cosmos in Native American Folklore.* University of New Mexico Press, Albuquerque.

# Index

# About the Author

The author grew up in Burlington, North Carolina, but spent many nights in remote backcountry campsites in the Uwharrie and Appalachian Mountains with fellow Boy Scouts and other friends. Since then he has been fortunate enough to spend nights under the stars in many remote locations, including the White Mountains of New Hampshire, the Sierra Nevada Mountains of southern California, on Mount Rainier in Washington State, and in the Brooks Range in Alaska.

After earning a Civil Engineering degree and a Professional Engineer's license, he combined his curiosity and a desire to create with his organizational skills to write a trail guide book for the Uwharrie Lakes Region in central NC. Star Trails: Navajo is his second book.

In addition to tenting under the stars, he also enjoys running, backpacking, mountain biking, canoeing, sea kayaking, and rock climbing. Most recently, adventure racing has allowed him to bring skills from many of these outdoor activities together with navigation challenges in a team challenge environment.

He also enjoys the company of a special group of friends known as the Adventure Family, with whom he often escapes from the world of work and "civilization" to enjoy the simple beauty of great camaraderie, great food, and amazing wilderness locations.

When not out adventuring in the wilderness, he can usually be found in front of a computer working for a living, or planning more trips.

The author lives in Cary, NC with his wife Sandy, and daughters Lainey and Lauren.